AMERICA'S FINEST ROCKETS™ PRESENTS:

ATLAS 116D

Atlas 116D - Publication Statement

Source Documents

The contents of this book are available for free online. This book is a compilation of reprints of several National Aeronautics Air and Space Administration (NASA) publications and figures, all relating to Atlas 116D. They were downloaded from online sources, including the NASA Technical Reports Server (NTRS, online at www.sti.nasa.gov) and Scribd.com.

Credits

The authors of the reports are generally attributable to the National Aeronautics and Space Administration (NASA) and the Lewis Research Center but, in some instances, authors of specific sections are listed within a report. The pictures on the cover are courtesy NASA.

Intent

Some readers do not like to read on a computer screen and prefer reading a physical copy. However, they may not want to suffer the price and inconvenience of a pile of unbound pages. This printed and bound copy of this report is priced low, to make it available to as many readers as possible. The reprinting of this report is not intended to offend NASA or the authors of the report, any of whom should contact the publisher with any questions or concerns.

Alterations and New Content

Nothing has been omitted from the original pdf files. The pages in the reports and articles have been renumbered for publication purposes, but no other changes have been made. However, some New Content has been added to this compilation: (1) the cover; (2) the preceding title page; (3) this Publication Statement; and (4) the Summary of the contents of this compilation on page (i).

Copyright

No copyright is asserted or claimed by the publisher for the *original* interior pages as downloaded. However, this compilation is ©2018 Mooncat® Collectibles. The original pages have been corrected using image processing software and are thus derivative works that are ©2018 Mooncat® Collectibles. The following are also ©2018 Mooncat® Collectibles: (a) the New Content; (b) the America's Finest Rockets logo and trademark; and (c) the Mooncat® logo and trademark. All rights reserved.

ISBN-13: 978-1544028613

America's Finest Rockets™

AMERICA'S FINEST ROCKETS™ PRESENTS:

ATLAS 116D

SUMMARY

Atlas 116D was never launched. It was used for dynamic testing of various configurations of Atlas-Centaur as the launch vehicle for Project Surveyor. During testing, water and styrofoam balls were used to simulate the weight of fuel in the rocket.

CONTENTS

ATLAS-CENTAUR-SURVEYOR LONGITUDINAL DYNAMICS TESTS

By Theodore F. Gerus, John A. Housely, and George Kusic

Lewis Research Center
Cleveland, Ohio

NATIONAL AERONAUTICS AND SPACE ADMINISTRATION

For sale by the Clearinghouse for Federal Scientific and Technical Information
Springfield, Virginia 22151 – CFSTI price $3.00

Atlas 116D arriving at Plum Brook on July 31, 1963. Image courtesy of Defense Video Imagery Distribution System (DVIDS).

ATLAS-CENTAUR-SURVEYOR LONGITUDINAL DYNAMICS TESTS

by Theodore F. Gerus, John A. Housely, and George Kusic

Lewis Research Center

SUMMARY

Full-scale dynamics tests were conducted at the NASA Lewis Research Center to investigate engine-structure coupled longitudinal oscillations (POGO) of the Atlas-Centaur launch vehicle. Dynamic characteristics determined are natural frequencies, damping ratios, mode shapes, and propellant pressure-force responses.

Experimental values of first mode natural frequencies were in good agreement with analytical values generated by General Dynamics Convair and Rocketdyne Division of North American Aviation, Incorporated. Second mode values were in close agreement for early flight times (near-full tanking condition) but theoretical values were much higher for near-empty tanking conditions. Mode shapes showed reasonably good agreement with the analytical model for this kind of experiment. The greatest discrepancy was found in the modal amplitudes of the propellant masses and propellant pressure-force responses. Structural damping values obtained showed that theoretical minimum requirements for stability of the engine-pump-structure loop are met in all cases, although marginal just before booster engine cutoff. Flight data have confirmed this marginal stability.

INTRODUCTION

Titan 11, Thor, and Jupiter vehicles have exhibited longitudinal oscillations (POGO) because of either dynamic characteristics of the engines and propellant feed systems alone or the coupling of engine and structural dynamics. The mechanism for the latter type of disturbance is basically a feedback system involving engine thrust perturbations which excite longitudinal structural modes and cause changes in propellant feed pressure which in turn cause continued thrust perturbations. The following sketch illustrates this loop phenomenon:

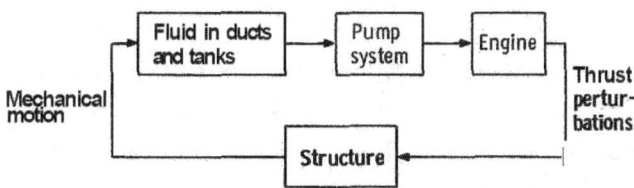

The loop dynamics must meet minimum stability requirements (attenuation and phase shift) in order to avoid sustained oscillations. These oscillations, if of a large magnitude or long duration, could result in engine shutdown and/or destruction of the vehicle.

Although previous Atlas flights have not encountered such phenomena, the addition of the Centaur upper stage changes the structural characteristics considerably. At the request of the NASA Lewis Research Center, analytical studies were conducted by General Dynamics/Convair (ref. 1) and by the Rocketdyne Division of North American Aviation, Incorporated (ref. 2) to investigate the possibility of these oscillations occurring in the Atlas-Centaur vehicle. These studies, based upon theoretical dynamic characteristics of the Atlas-Centaur, indicated that a relatively high degree of structural damping was needed to maintain stability. In an attempt to experimentally verify these analyses, a series of longitudinal dynamics tests on a full-scale Atlas-Centaur vehicle were conducted in a dynamic test facility at the Plum Brook Station of Lewis from November, 1963 to July, 1964. Natural frequencies, mode shapes, damping, and propellant pressure responses were studied. Particular emphasis was placed on determining the characteristics of the liquid oxygen (lox) supply system rather than fuel system, since early analytical studies showed that the lox supply system was the major contributor to POGO instability.

TEST SETUP

The Plum Brook E-site (fig. 1) which had been originally built to test an Atlas vehicle was heightened to accommodate the Atlas-Centaur combination. Test hardware for the first nine tests consisted of the 116-D Atlas, a flight-type interstage adapter, and a rigid water-filled tank (dummy Centaur) having the same mass as the Centaur-Surveyor. The use of a dummy mass was used because of the unavailability of a flight-type Centaur tank in the desired time period. Justification for this decision was based upon an early General Dynamics/Convair study of the Atlas-Centaur structural response which indicated that the upper stage would be moving essentially as a rigid body in the first two longitudinal modes at flight times of interest. When a Centaur tank, insulation panels, nose fairing, and Surveyor model were received, tests 10 to 15 were run with these in place of the dummy Centaur.

4

To facilitate mating with the suspension system, the Atlas was modified by replacing all structure and components aft of the thrust barrel with an equivalent mass I-beam structure X-frame. A suspender consisting of a steel cable, spring box, hydraulic cylinder, and a load cell was fastened to each of the four frame ends (fig. 2). Each spring box (fig. 3), contained 4 to 16 springs, with a constant of about 400 pounds per inch (7.0×10^4 N/m) per spring. The number of springs for each test was proportioned to the vehicle weight of the particular configuration, in order to give a static deflection of 1 foot (0.3 m), thus keeping the natural frequency of the suspension system well below the range of vehicle resonances. Lateral stability of the system was provided at the bottom by the $1/2^o$ (8.7×10^{-3} rad) inclination of the cables and at the top by horizontal springs (fig. 2).

In order to avoid operational problems involved with the handling of cryogenic propellants, an equal volume of deionized water was tanked instead of lox and RP fuel, and polystyrene balls having an equivalent bulk density replaced the liquid hydrogen. For testing, the propellant tanks were maintained at flight pressures of 29.5 and 59.0 pounds per square inch gage (2.03×10^5 and 4.06×10^5 N/m^2 gage) in the Atlas lox and fuel tanks, respectively, and 15 and 5 pounds per square inch gage (1.03×10^5 and 3.44×10^4 N/m^2 gage) in the Centaur lox and liquid hydrogen tanks. Inasmuch as a hazardous condition with respect to tank rupture exists at these pressures, it was necessary to control all operations remotely, once the tanking procedures were begun. These remote operations were conducted from a control room (H-building) which was approximately 1/4 mile (400 m) away. Television cameras were used to monitor the vehicle.

INSTRUMENTATION

Instrumentation for tests 1 to 9 (dummy Centaur) consisted of strain-gage-type accelerometers and pressure transducers. Locations on X and Y axes are shown in figure 4. For tests 10 to 15 (actual Centaur) the accelerometers were relocated to cover the Centaur stage and foil-type strain gages were mounted on the Atlas tank skin. Locations are shown in figure 5. Pressures were measured in the lox system by close-coupled strain-gage-type transducers located in the lox system as shown in figure 6. Load cells were used for both vehicle weighing and a measurement of driving force level. All modal data were digitally recorded and reduced with a digital computer program. Sixteen channels of data which were randomly selected were recorded on oscillograph paper. All damping was determined through the use of analog oscillograph paper. End-to-end system accuracy of the instruments is considered to be about 2 percent of full scale. Full-scale ranges of the instruments were as follows:

Accelerometers, g (m/sec^2) . . . . ± 1.0 (± 9.8)

Load cell (force), lb (N) . . . . ± 1000 (4.45)

Pressure, psia (N/m^2 abs) . . . . 0 to 100 (0 to 6.89×10^3)

TEST PROCEDURE

For testing, the Atlas tanks were filled to a level representing 0, 30, 60, 90, 120, 132, 144, or 151 (BECO) seconds of flight time and pressurized to flight pressures. Table I describes the test configurations. Excitation was then applied to the suspended vehicle by an electrodynamic shaker via load cell and X-frame at frequencies varying from 6 to 40 hertz. Input force levels ranged from 500 to 5000 pounds (2.22×10^2 to 2.22×10^3 N) with resulting accelerations of 0.6 g (5.9 m/sec^2) zero-to-peak maximum on the vehicle.

When resonant conditions were determined, transducer output was recorded on analog recorders and on digital tape. Data were taken at discrete points near each resonance peak to define the response curve. At the resonance peak the shaker was electrically decoupled allowing natural decay of the oscillations with transducer output being recorded on the analog recorder. This procedure was followed to identify the first two modes of each tanking condition.

RESULTS AND DISCUSSION

The values of natural frequency obtained in tests made with the dummy Centaur were significantly different from those found in tests with the actual flight-type stage. Although data from both series of tests are presented for comparitive purposes, the discussion is limited to the comparison of experimental results with theoretical analysis using the Centaur tank and SD-4 dummy Surveyor.

Analytical values of natural frequencies and mode shapes are taken from the latest studies performed by General Dynamics/Convair (ref. 1). The theoretical maximum values of pressure-force response reported herein are based on the block diagram of appendix A and make use of analytical mode shapes and modal masses and experimentally determined values of damping.

6

Natural Frequencies

Figure 7 compares theoretical and experimental natural frequencies of the Atlas-Centaur-Surveyor for the first two longitudinal modes. The first modal frequencies appear to agree very well but divergence between the second mode predicted and measured frequencies indicate some analysis limitations. The second mode frequencies agree very well at the time of predicted second mode POGO (30-sec flight time). Also shown in figure 7 are experimental Atlas-dummy Centaur data, illustrating the inaccuracies in natural frequencies resulting from assuming Centaur a rigid body.

The first mode resonance at 90 seconds and the second mode resonance at 151 seconds were not found. In the first case it is believed that coincidence of a node with the point of force input prohibited modal excitation. In the latter case an area of weak vehicle response was noted at about $16\frac{1}{2}$ hertz which was probably the second mode, but the transducer signals were of such low strength and poor quality that the actual peak response could not be accurately determined.

Mode Shapes

Relative movement of vehicle segments was determined directly from accelerometer outputs except for lox and fuel mass motions which were determined from strain gage data by the method of appendix B. A comparison of modal amplitudes for the first two longitudinal modes is shown in tables II and 111. It should be noted here that the amplitudes represented are the vector sum of real and imaginary components with the sign determined by the direction of the real component relative to the maximum amplitude, arbitrarily set at $+10$. Although most data had very small imaginary amplitudes (relative to maximum amplitude), no attempt will be given in this report to separate the real and imaginary terms.

The differences between theoretical and experimental modal deflections are smallest for the first mode early flight times and largest for second mode late flight times. This is common with analytical modeling of structural modes. The following is a list of possible sources of differences between the analytical and experimental mode shapes (no relative significance is given here for these suggested differences):

(1)Representation of fuel and lox as single masses with skins deflected as a truncated cone. Strain gage data indicate this assumption fair for lower modes but questionable for higher modes. Figures 8 and 9 show the hoop strain resulting from longitudinal oscillations,

7

(2) Measurement and computation accuracy. Since fuel and lox modal deflection are computed by using hoop strain gages with assumptions as given in appendix B, some limitation in accuracy is probable. Other normal instrument accuracies as given in the instrument section may also contribute.

(3) Finite mass and spring assumptions. Although it is customary to break a continuous system into finite springs and masses, the degree of accuracy is always limited by the choice of the separation. Figures **10** and **11** show outputs of accelerometers mounted on the tank skin for each mode which show continuous motions of masses. It should be noted that these are plotted in a lateral plane to show longitudinal acceleration.

(4) Normal mode assumptions in analytical model. In the analytical model there are no imaginary vectors when the mass modal deflections are computed. It is generally impossible to excite a true normal mode experimentally, especially with a single force exciter.

(5) Suspension system limitations. Although a great deal of care was taken in the design of the suspension system, it is not a true free-free system.

(6) Nonlinearities and Poisson's ratio effect. All equations in the analysis are assumed linear. Damping data in the next section show these limitations. Poisson's ratio is not included in the analysis.

Damping

Standard technique of analysis of decay curves (of which fig. **12** illustrates several typical) shows that the decay is not exponential and that both viscous and coulomb damping are present. Although the concept of critical damping is based on a viscously damped second order system, an equivalent value for this system may be determined by basing the logarithmic decay computation on the number of cycles required to decay from maximum to near zero amplitude. Damping ratios obtained by this method meet theoretical minimum requirements for stability in all cases. It is interesting to note that first mode small amplitude POGO oscillations (0.125-g (**1.23** m/sec^2) zero-peak (ref. **3**)) occurred at the time when stability margins are smallest, but second mode POGO occurred when analysis indicated a high stability margin. This indicates that the second mode accuracy is less than that of the first mode. Data shown in figure **13** indicate damping measured in Atlas-Centaur tests ranging from **1.1** percent of critical to **4.8** percent of critical at liftoff and Booster Engine Cutoff, respectively. Missing data points occur where either no mode was found or where the decay curve was of poor quality. Figure **14** indicates damping measured in Atlas-dummy Centaur tests for comparitive purposes.

Evidence that the damping characteristics may be undergoing a softening characteristic with force (less damping with greater amplitude, ref. **4**) is indicated by the shifts **in**

8

peaks in figure 15. Here it will be noticed that an increase in force causes a greater than proportional increase in amplitude and that the frequency of peak response is lowered. The range of accelerometer response for this data is from 0 to 0.5 g (4.9 m/sec^2) in comparison with the 0.125 g (1.23 m/sec^2) seen in flight. It was suggested that the source of this effect might be the polystyrene balls used in the Centaur hydrogen tank and/or Centaur insulation panels, but an additional test run with the tank empty and no insulation panels showed insignificant difference in damping values.

Lox Duct Pressure Responses

Transducers mounted in the main lox line and at the capped ends of pump feed lines as shown in figure 6 were used to determine pressure/force responses. Although data were obtained for all cases, only one portion of the flight is shown here, arbitrarily chosen as 60-second flight time. The analytical data shown were based upon theoretical mode shapes and experimentally measured damping (appendix B). The pressure responses were computed for a nonflowing liquid, specifically the E-stand configuration. Figures 16 to 19 show analytical values much higher than those measured in the tests. Furthermore, the sustainer and booster lox pump inlet pressures measured show a rise in pump inlet pressures below structural responses. These data are difficult to interpret since the lines could not be mechanically terminated in the same way as they are in flight. Part of the difference in response could be attributed strictly to unrealistic motion of the pipes themselves. Experimental and theoretical values for the 60-second flight time (figs. 16 to 19) show the analytical values to be orders of magnitude higher in the first mode and by as much as 20 to 1 in the second mode. These discrepancies were typical for all tests.

With an exciting frequency near 6 hertz, high pressure was noted in the lox feed lines. This is attributed to a pipe resonance since the frequency did not coincide with a vehicle natural frequency and did not occur in the lox tank bottom.

Theoretical data presented for comparison are based on the Plum Brook configuration (i.e., no engines, no flow, and water in tanks). A block diagram describing the various dynamic terms for the configuration is presented in appendix B.

CONCLUSIONS

The test data presented show a close correlation with predicted frequencies and a reasonable correlation with mode shapes for this kind of experiment. Although the analytical models used for comparison were the most recent models generated by General

Dynamics/Convair, upgrading is continuing in an effort to make the structural analytical model as accurate as possible. With the present analytical model of the structure, measured damping, and calculated line and engine responses, it is possible to generate a stability limit that closely agrees with flight data in the first mode. However, since no reasonable measurements have been made coupling fluid lines and pumps, it is felt that any further test activity should be in this direction.

Lewis Research Center,
National Aeronautics and Space Administration,
Cleveland, Ohio, June 15, 1967,
491-05-00-01-22.

APPENDIX A

LOX TANK PRESSURE SYSTEM E-STAND CONFIGURATION

 The following block diagram is the analytical model of the part of the loop used for
E-stand test configuration.   This is part of the loop used in the complete system
ytical model and modified for use here.   P* is the pressure derived from the force
sidering the sum of the structural response of the aft end of the vehicle and the lox
s. P_J, P_1, P_2, and P_s are pressures computed utilizing P* and the response of
ducts.

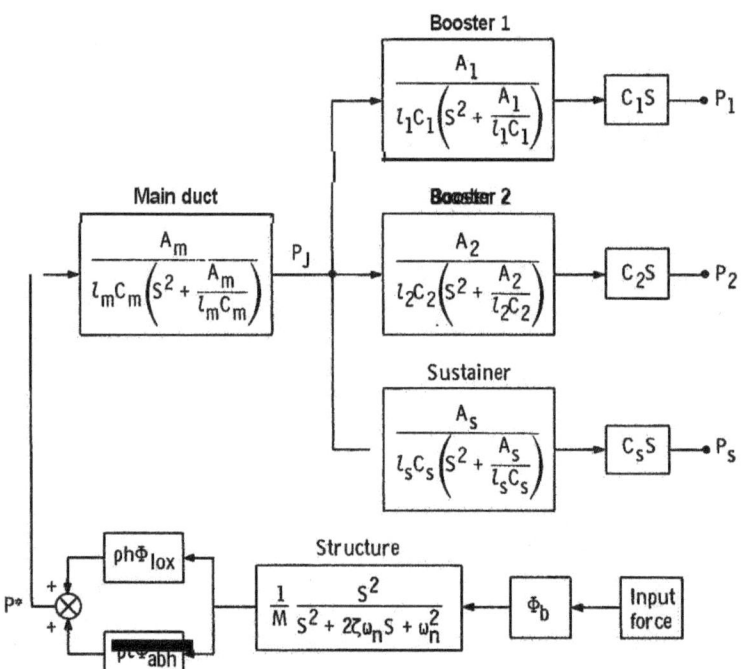

APPENDIX B

METHOD FOR COMPUTING FLUID MASS MOTION FROM STRAIN GAGE DATA

This analysis assumes that the **tank** bottom is rigid and that strain between gages has a linear distribution. As shown in the sketch at the left, longitudinal oscillations will

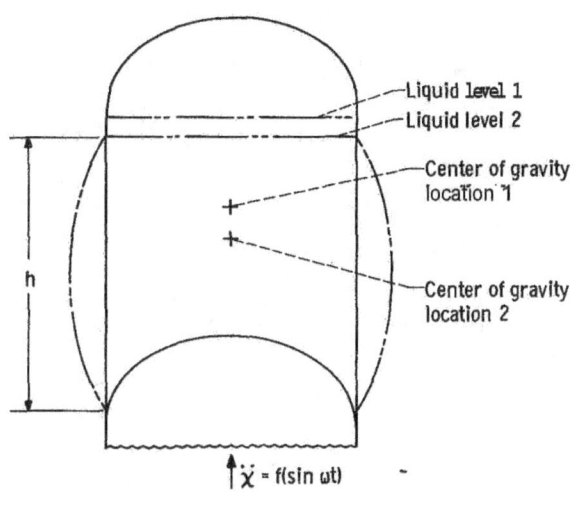

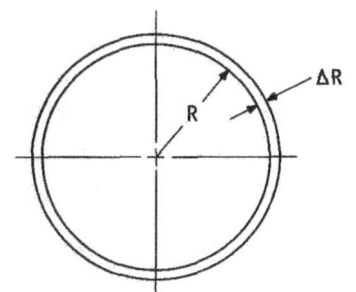

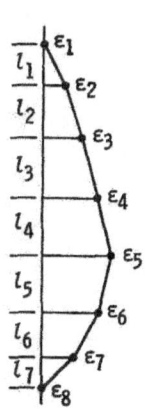

cause a change in tank volume, due to pressure fluctuation on the **tank** wall as given by $\Delta P = \ddot{\chi} h \rho$ where $\ddot{\chi}$ is the induced acceleration, h is the height of liquid, and p is the density of liquid. (Symbols are defined in appendix C.) Strain gages located on the **tank** skin may be used to compute the resulting motion of the liquid center of gravity. The change in area of any section is given with good accuracy by

$$AA = 2\pi R(\Delta R)$$

where AR is a function of hoop strain (see sketch at left) and is given by

$$AR = \frac{A \text{ circumference}}{2s} = \frac{\epsilon(2\pi R)}{2a}$$

Where ϵ is unit strain, then

$$AA = R\epsilon(2\pi R)$$

With a pressure distribution as shown in the sketch at the left where ϵ_1, ϵ_2, $\cdots$, ϵ_8 are measured hoop strains, the volume change is given by

$$\Delta V = 2\pi R^2 \sum_{n=1}^{n=7} \left(\epsilon_n + \frac{\epsilon_{n+1} - \epsilon_n}{2} \right) l_n \qquad (A1)$$

The change in height of liquid will then be given by

$$\Delta h = \frac{\Delta V}{\pi R^2}$$

and the change in center of gravity location is $\beta \Delta h$ where β is a coefficient which considers tank geometry such as dome or cone bottoms. The sensitivity of β to liquid level for the Atlas tanks is shown in the sketch at the left. When sinusoidal motion is assumed, the acceleration with respect to the tank bottom is

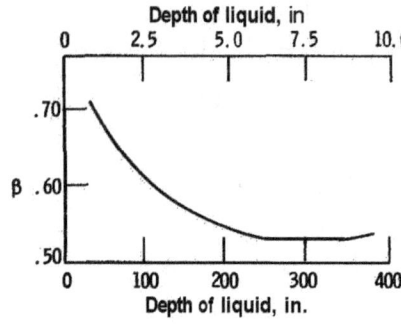

$$|\ddot{x}| = \beta \frac{\Delta V}{\pi R^2} \omega^2 \tag{A2}$$

where ω is the circular frequency. Combining equations (A1) and (A2) results in

$$|\ddot{x}| = 2\beta\omega^2 \sum_{n=1}^{n=7} \left(\epsilon_n + \frac{\epsilon_{n+1} - \epsilon_n}{2}\right)l_n \tag{A3}$$

The total acceleration of the center of gravity is then the sum of equation (A3) and the acceleration of the tank bottom.

APPENDIX C

SYMBOLS

A   cross sectional area of **lox** duct, in.^{2} (m^2)

C   capacitance of duct, in. -sec^2 (m-sec^2)

h   depth of liquid in Atlas lox tank, in. (m)

l length of duct, in. (m)

M   generalized mass of structural mode being considered, (lb)(sec^2)/in. (kg)

P   pressure at pump inlet, psi (N/m^2)

P*   pressure at top of main duct, psi (N/m^2)

P_J pressure at bottom of main duct, psi (N/m^2)

R   radius of tank, in. (m)

S Laplacian operator, sec^{-1}

V   volume of tank, in.^{3} (m^3)

E unit strain, in. (m)

ζ damping ratio, dimensionless

P density of water, lb/in.^{3} (N/m^3)

Φ_{abh} modal amplitude of aft bulkhead, dimensionless

Φ_b modal amplitude of gimbal plane, dimensionless

Φ_{lox} modal amplitude of lox mass, dimensionless

$\ddot{\chi}$ induced acceleration, g (m/sec^2)

ω circular frequency, rad/sec

ω_n undamped natural frequency, rad/sec

Subscripts:

m   main lox duct

s   sustainer engine lox duct

1   engine 1 booster lox duct

2   engine 2 booster **lox** duct

14

REFERENCES

1. Rose, Robert G. ; Simson, Anton K. ; and Staley, James A. : A Study of System-Coupled Longitudinal Instabilities in Liquid Rockets. Part 1: Analytic Model. Rep. No. GD/C-DDE65-049, pt. 1 (AFRPL-TR-65-163, pt. 1, DDC No. AD-471523), General Dynamics/Convair, Sept. 1965.

2. Wolf, K. E.; Austin, E. A. and Nelson, R. L. : Study of Longitudinal Oscillations During Flights of Atlas Space Launch Vehicles. Rep. No. AER 64-2, Rocketdyne Div., North American Aviation, Inc., Mar. 30, 1964.

3. Staff of Lewis Research Center: Post Flight Evaluation of Atlas-Centaur AC-4 (Launched December 11, 1964). NASA TM X-1108, 1965.

4. Harris, Cyril M. ; and Crede, Charles E., eds. : Shock and Vibration Handbook Vol. 1. McGraw-Hill Book Co., 1961.

TABLE I. - **TEST** CONFIGURATIONS

Test	Flight, time, sec	Volume of water in Atlas fuel tank, V_F		Volume of water in Atlas lox tank, V_L		Configuration	Total test weight	
		ft3	m³	ft3	m³		lb	kg
1	-----	0	0	0	0		29 400	13 336
2	151.7	151	4.28	221	6.26		67 990	30 840
3	144	197	5. 58	298	8.43		75 670	34 323
4	132	307	6.69	461	13.6		93 950	42 615
5	120	416	11.6	663	18.8		112 110	50 852
6	90	723	20.5	1113	31.5		159 350	72 280
7	60	958	27.1	1562	44.2		202 030	91 639
8	30	1227	34.7	2015	57.1		247 270	112 159
9	0	1494	42.3	2473	70.0		292 320	132 593
10	60	956	27. 1	1562	44.2		202 030	91 639
11	151.7	151	4.28	221	6.26		67 990	30 840
12	144	197	5.58	298	8.43		75 670	34 323
13	0	1494	42.3	2473	70.0		292 320	132 593
14	90	723	20.5	1113	31. 5		159 350	72 280
[b]15	151.7	151	4.28	221	6.26		66 610	30 214

[a]**Centaur** tanks (configuration 2) contained 4630 lb (2100 kg) of polystyrene balls (LH_2) and 310 cu ft (8.77 cu m) of water (LO_2).

[b]**Run** without insulation panels.

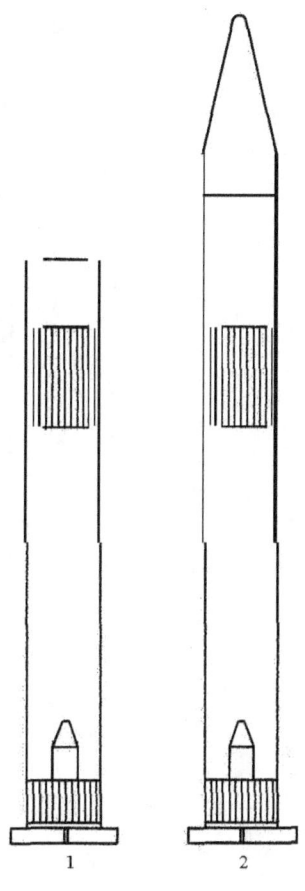

1   2

Configuration

16

TABLE II. - FIRST MODE RELATIVE AMPLITUDE

[No experimental data for 90-second flight time.]

Mass number	Flight time, sec							
	0		60		144		151	
	Theoretical	Experimental	Theoretical	Experimental	Theoretical	Experimental	Theoretical	Experimental
5	0.697	0.39	0.959	0.80	0.906	0.81	-0.752	0.82
7	.679	.40	.912	.86	.797	.75	-.643	-.73
9	.713	.42	1.00	.95	1.00	1.00	-.838	-1.00
11	.615	.31	.750	.66	.436	.31	-.300	-.28
12	----	----	.608	.62	-.365	-.21	.501	.36
13	-.836	-.14	-.928	-.058	-.926	-.73	1.00	.68
14	.687	.58	.526	.71	-.52	-.60	.615	.59
15	----	----	.589	.83	----	----	----	----
16	1.00	1.00	.763	.99	----	----	----	----
17	.829	.72	.679	.98	-.636	-.77	.741	.85
18	.840	.73	.694	1.00	-.672	-.79	.790	.88

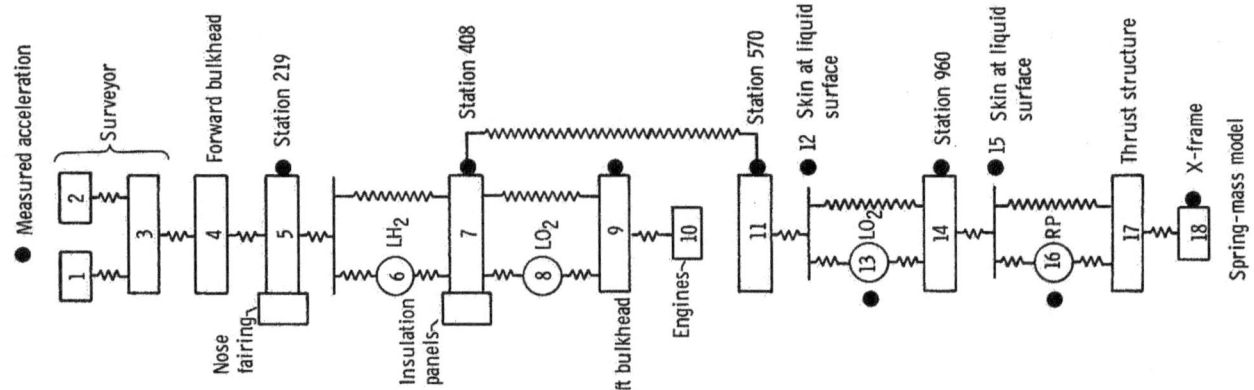

TABLE III. - SECOND MODE RELATIVE AMPLITUDE

Mass number	Theoreti-cal	Experi-mental	Theoreti-cal	Experi-mental	Theoreti-cal	Experi-mental	Theoreti-cal	Experi-mental
	0		60		90		144	
5	0.937	0.83	-0.926	-0.83	-0.794	-0.84	-0.95	-0.16
7	.865	.85	-.841	-.82	-.703	-.80	-.061	-.12
9	1.000	1.00	-1.00	-1.00	-.872	-1.00	-.121	-.12
11	.619	.56	.556	.49	-.402	-.44	.034	-.04
12	-----	-----	-.307	-.29	.001	-.06	.235	-.19
13	.038	-.13	-.045	-.02	-.296	-.05	.801	-.10
14	-.042	-.056	-.269	.22	.535	.44	-.410	.25
15	-----	-----	.359	.37	.768	.64	------	-----
16	-.336	-.077	-.616	.56	1.00	.76	------	-----
17	-.167	-.073	-.489	.48	.891	.96	.851	.98
18	-.172	-.085	.510	.49	.953	.96	1.00	1.00

18

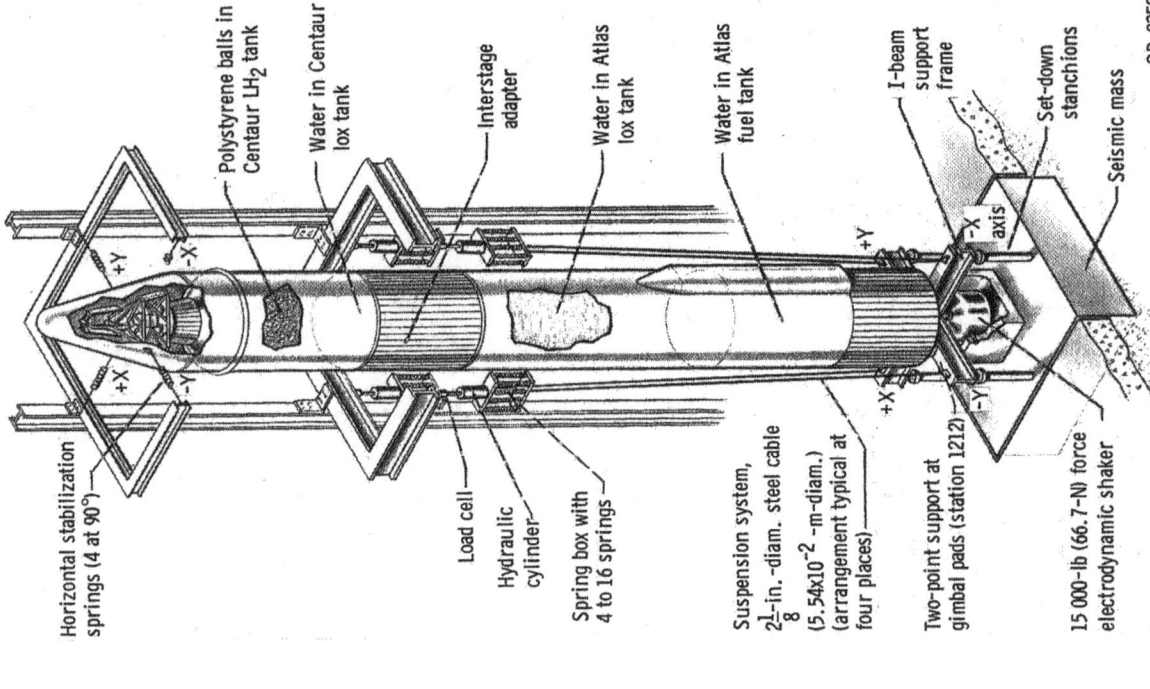

Polystyrene balls in Centaur LH$_2$ tank

Water in Centaur lox tank

Interstage adapter

Water in Atlas lox tank

Water in Atlas fuel tank

I-beam support frame

Set-down stanchions

Seismic mass

CD-8359

Horizontal stabilization springs (4 at 90°)

Load cell

Hydraulic cylinder

Spring box with 4 to 16 springs

Suspension system, $2\frac{1}{8}$-in.-diam. steel cable (5.54×10^{-2}-m-diam.) (arrangement typical at four places)

Two-point support at gimbal pads (station 1212)

15 000-lb (66.7-N) force electrodynamic shaker

+Y   -X'

+X   -Y'

+Y

-X axis

+X

-Y

Figure 2. – Vehicle support system.

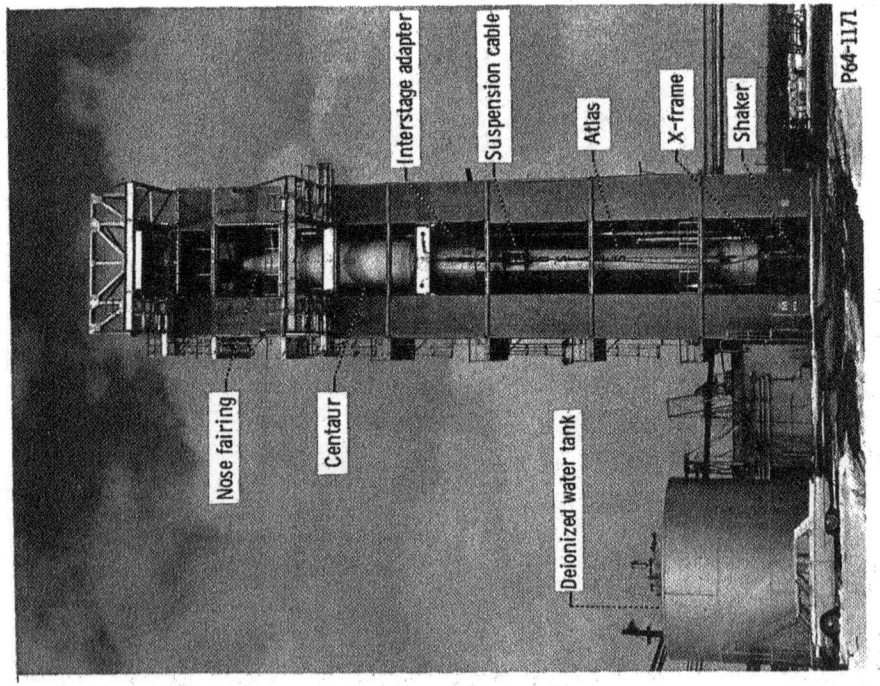

Nose fairing

Centaur

Interstage adapter

Suspension cable

Atlas

X-frame

Shaker

Deionized water tank

P64-1171

Figure 1. – E-stand test facility with Atlas-Centaur-Surveyor.

19

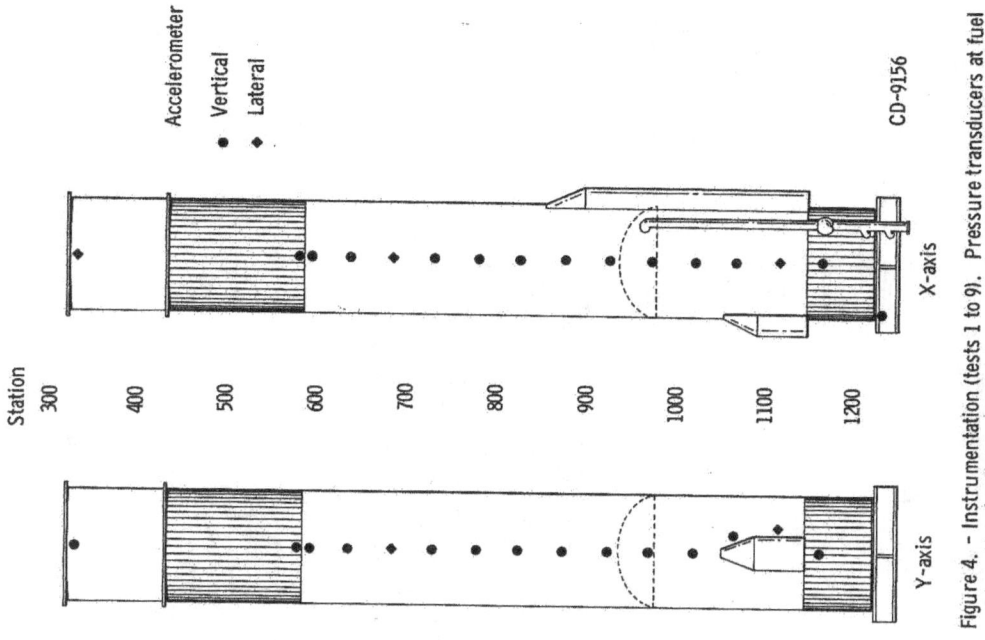

Station

Accelerometer

● Vertical
◆ Lateral

CD-9156

X-axis

Y-axis

Figure 4. - Instrumentation (tests 1 to 9). Pressure transducers at fuel cone apex, lox line butterfly valve, and lox line upper valve.

Load cell

Spring box

Suspension cable

P-64-1022

Figure 3. - Closeup of suspension springs.

20

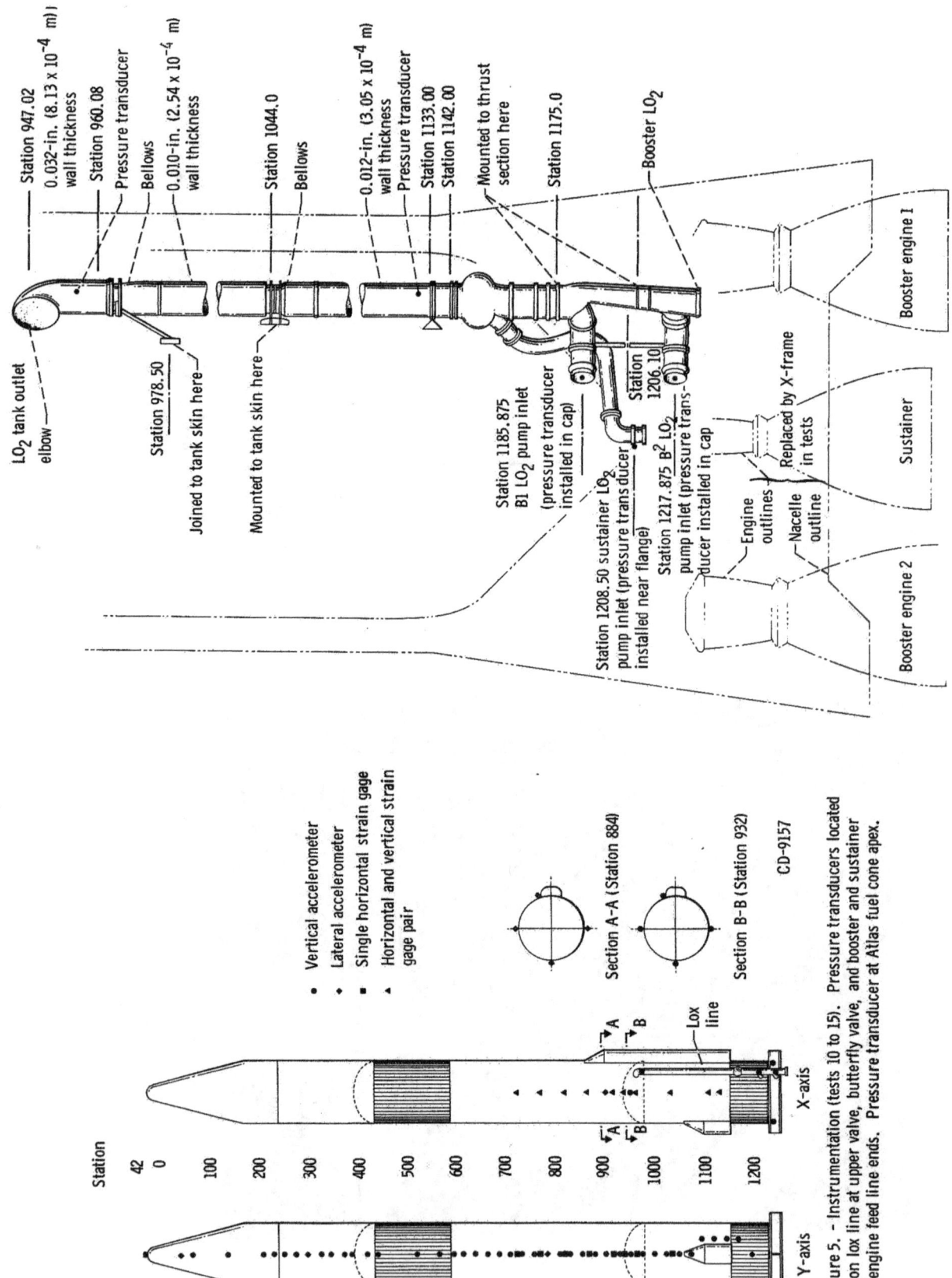

Station 947.02
0.032-in. (8.13 x 10⁻⁴ m) wall thickness
Station 960.08
Pressure transducer
Bellows
0.010-in. (2.54 x 10⁻⁴ m) wall thickness
Station 1044.0
Bellows
0.012-in. (3.05 x 10⁻⁴ m) wall thickness
Pressure transducer
Station 1133.00
Station 1142.00
Mounted to thrust section here
Station 1175.0
Booster LO₂

LO₂ tank outlet elbow
Station 978.50
Joined to tank skin here
Mounted to tank skin here

Station 1185.875 B1 LO₂ pump inlet (pressure transducer installed in cap)
Station 1208.50 sustainer LO₂ pump inlet (pressure transducer installed near flange)
Station 1206.10
Station 1217.875 B2 LO₂ pump inlet (pressure transducer installed in cap)

Booster engine I
Sustainer
Booster engine 2
Engine outlines
Nacelle outline
Replaced by X-frame in tests

CD-9158
and instrumentation used in tests.

Station
42
0
100
200
300
400
500
600
700
800
900
1000
1100
1200

• Vertical accelerometer
♦ Lateral accelerometer
■ Single horizontal strain gage
▲ Horizontal and vertical strain gage pair

Section A-A (Station 884)
Section B-B (Station 932)
CD-9157

A
B
Lox line
X-axis
Y-axis

Figure 5. - Instrumentation (tests 10 to 15). Pressure transducers located on lox line at upper valve, butterfly valve, and booster and sustainer engine feed line ends. Pressure transducer at Atlas fuel cone apex.

21

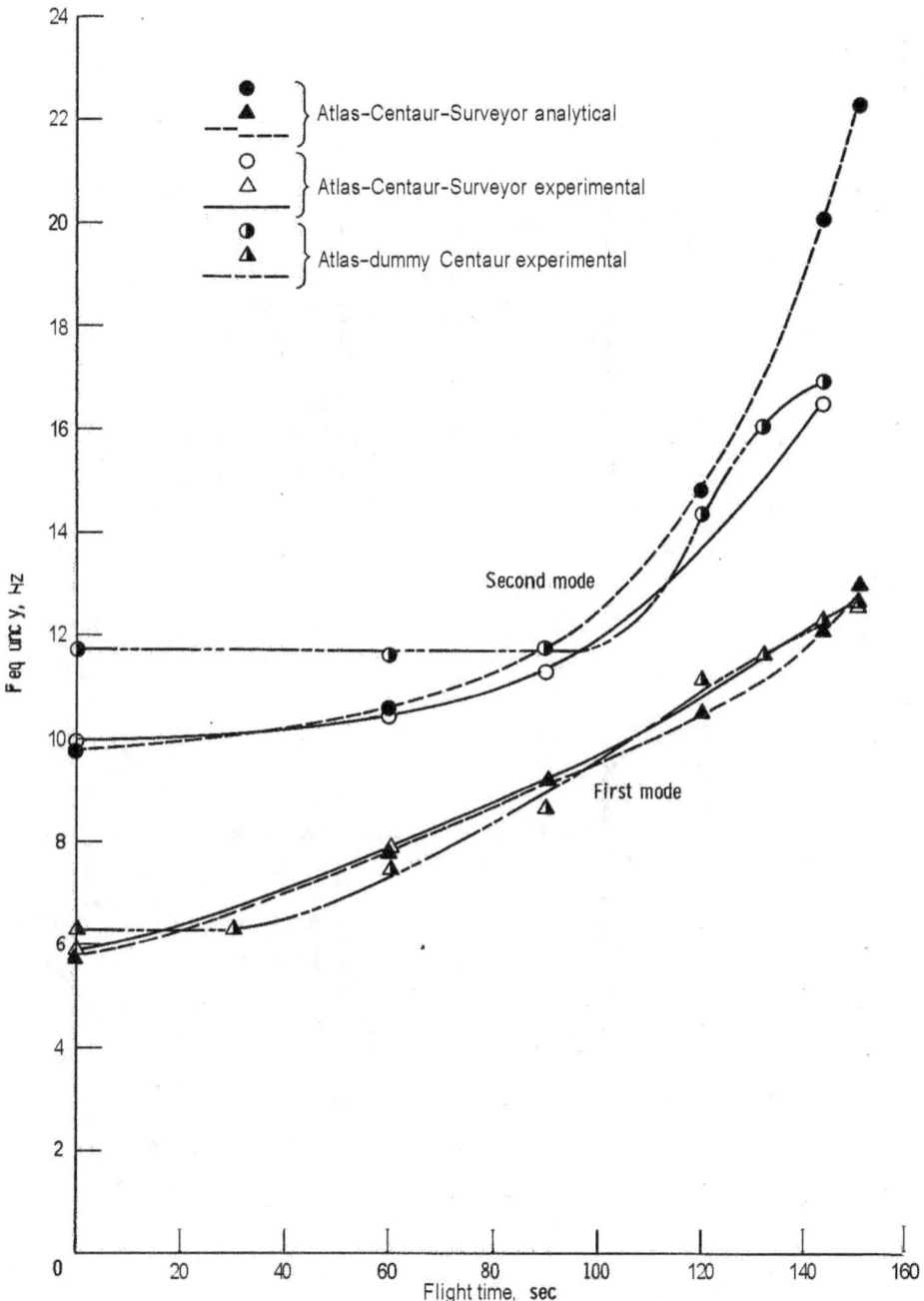

Figure 7. - Natural frequency plotted against flight time for Plum Brook configuration.

22

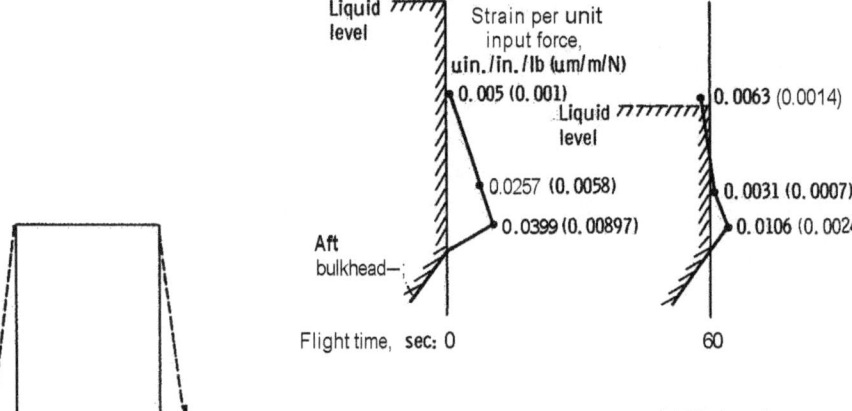

(a) First mode.

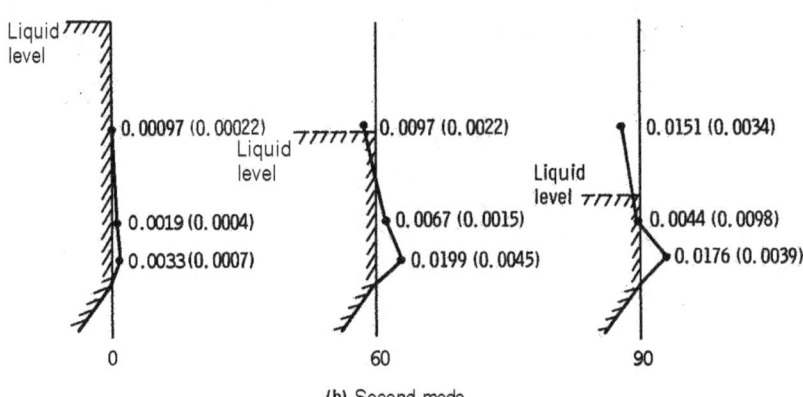

(b) Second mode.

Figure 8. - Atlas fuel tank hoop strain at resonance.

Spring-mass model
analysis assumes
that displaced tank
skin takes form of
truncated cone as
above

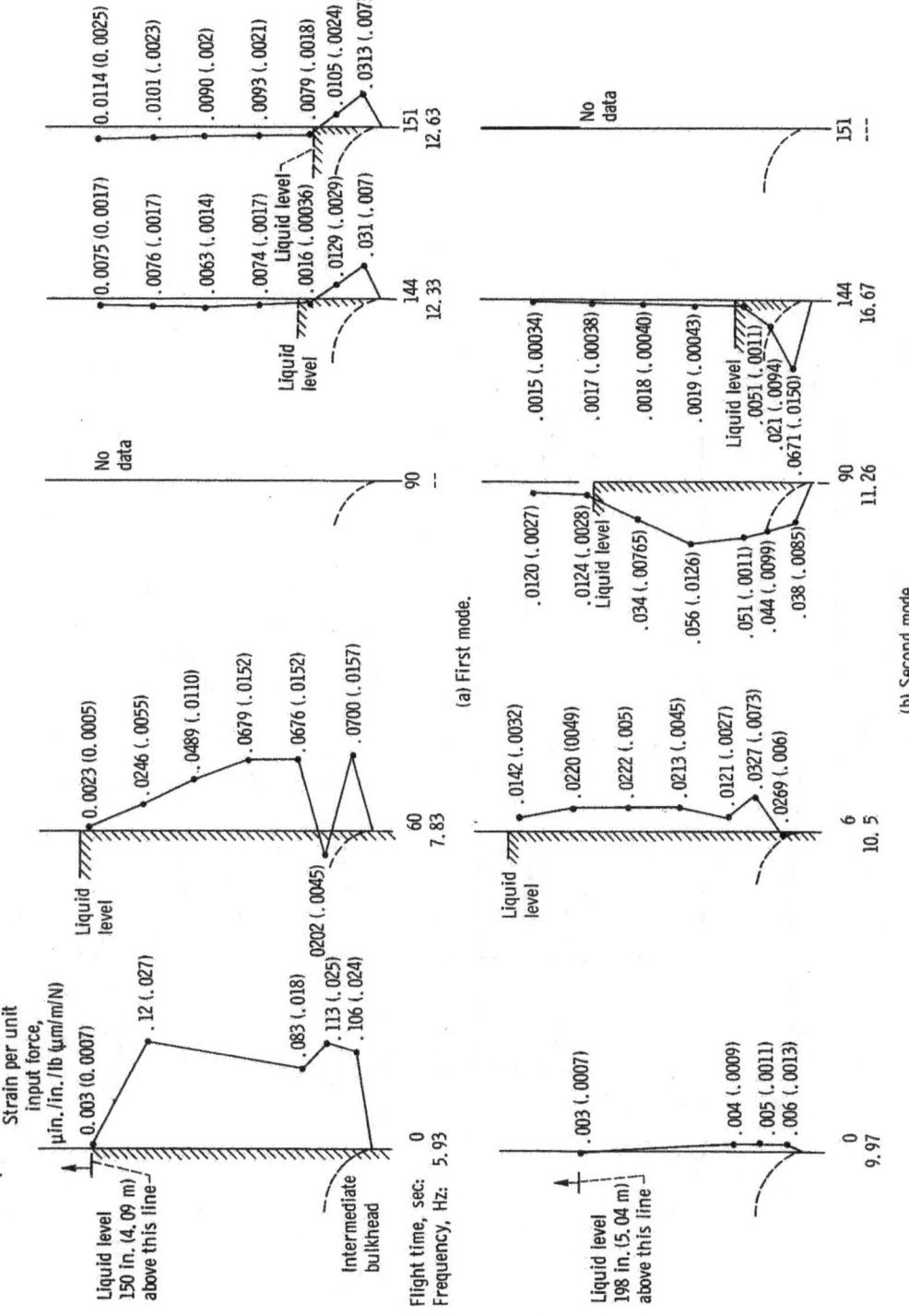

Figure 9. – Atlas lox tank hoop strain at resonance. (Vertical – not to scale).

(a) First mode.

(b) Second mode.

24

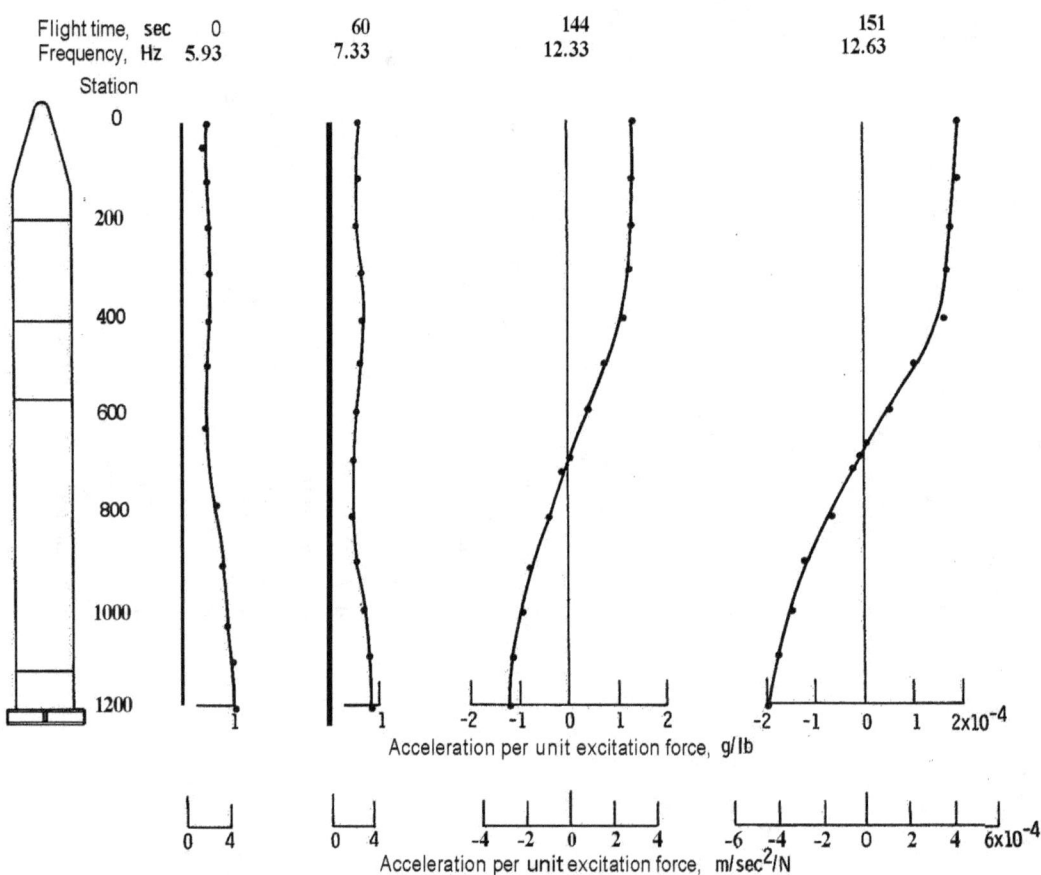

Figure 10. – First **mode** longitudinal acceleration plotted against station.

25

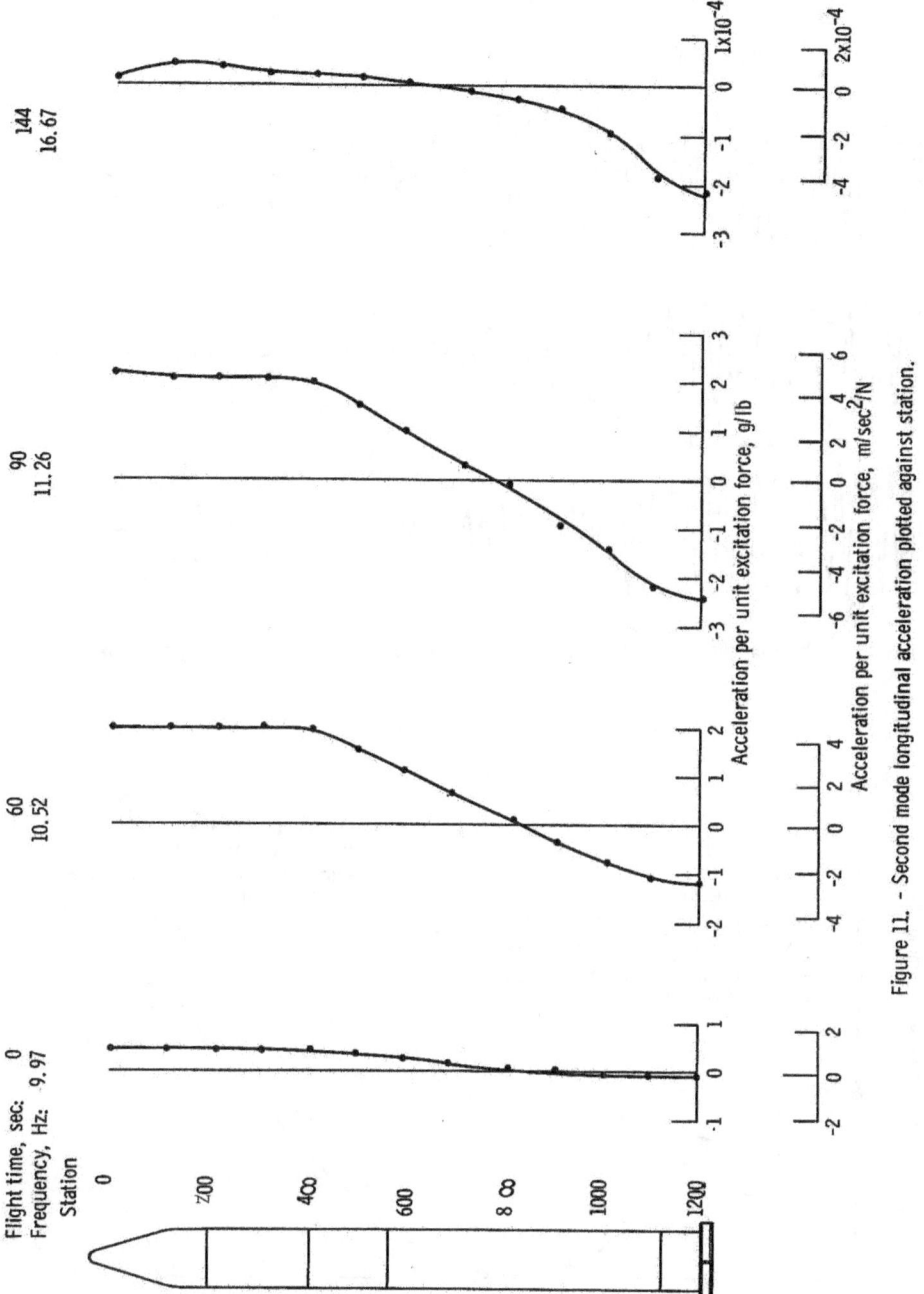

Figure 11. - Second mode longitudinal acceleration plotted against station.

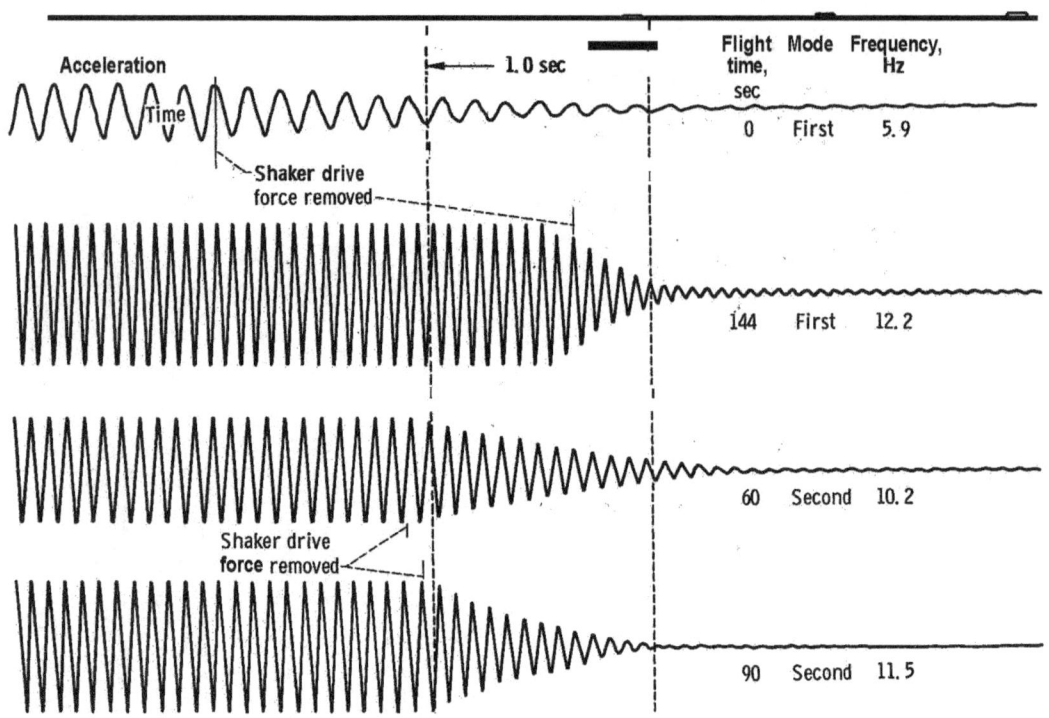

Figure 12. - Atlas-Centaur-Surveyor test stand typical data used to calculate damping by decay methods.

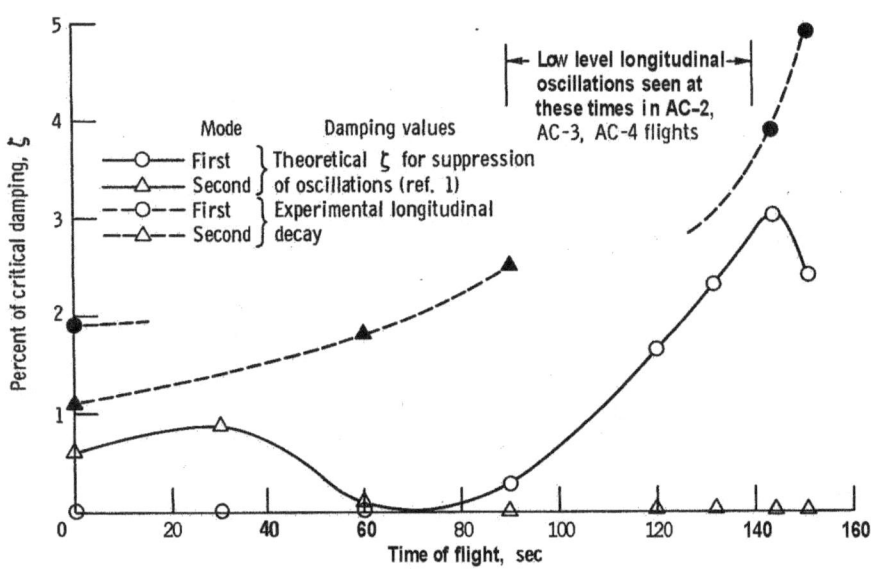

Figure 13. - Damping variations with flight time.

27

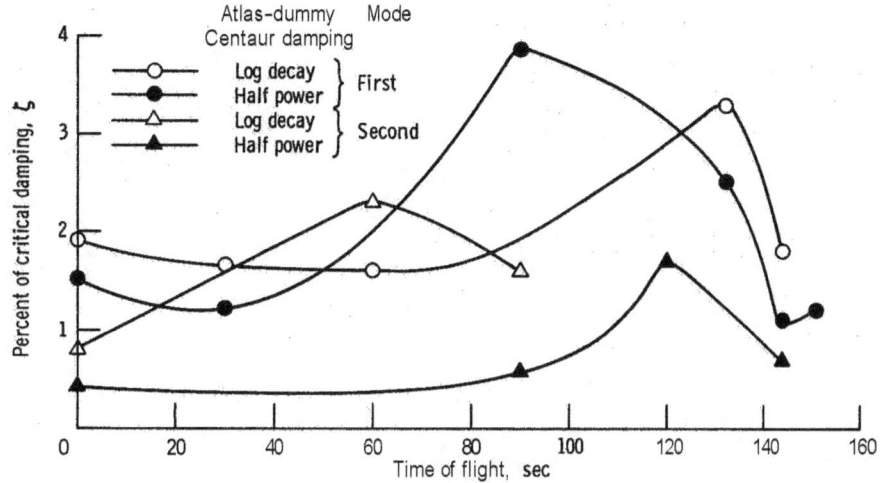

Figure 14. - Atlas-dummy Centaur damping variations with flight time.

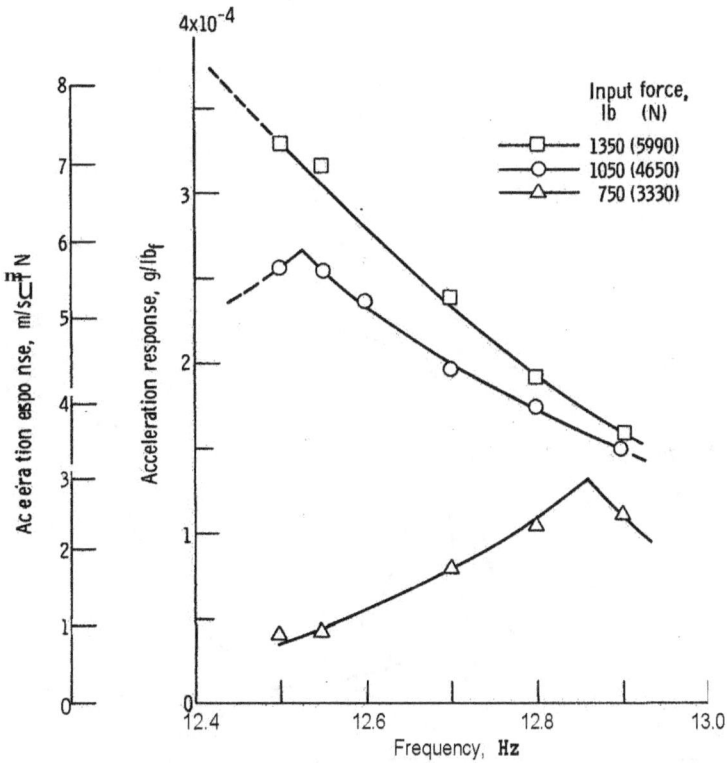

Figure 15. - First mode response. Test 11; booster engine cutoff condition (151 sec).

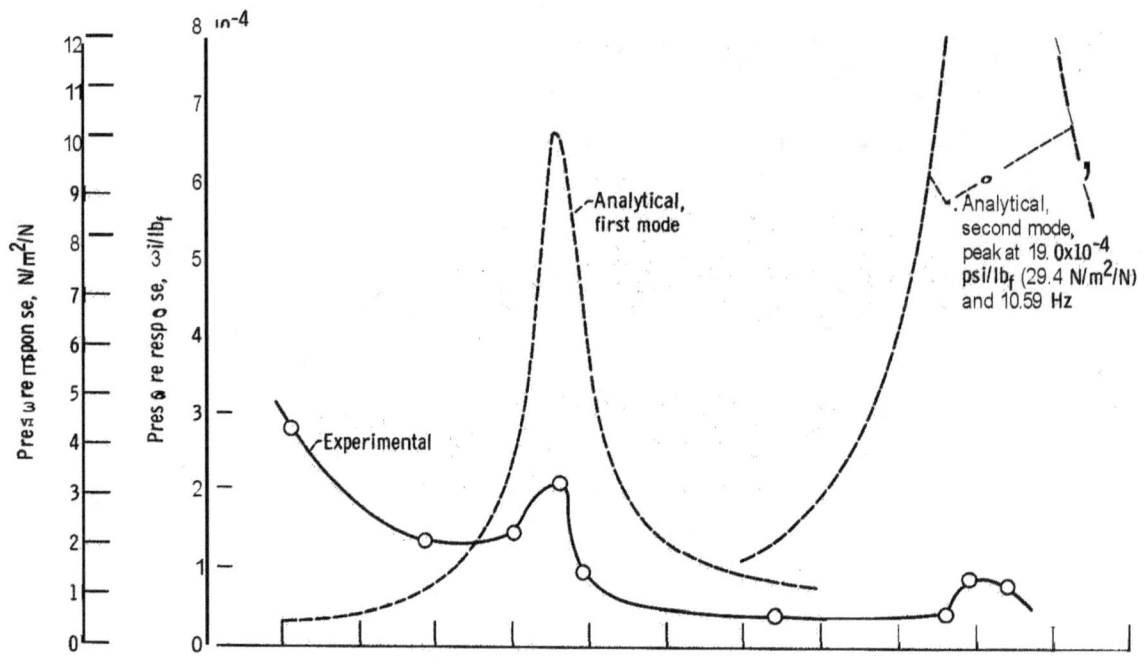

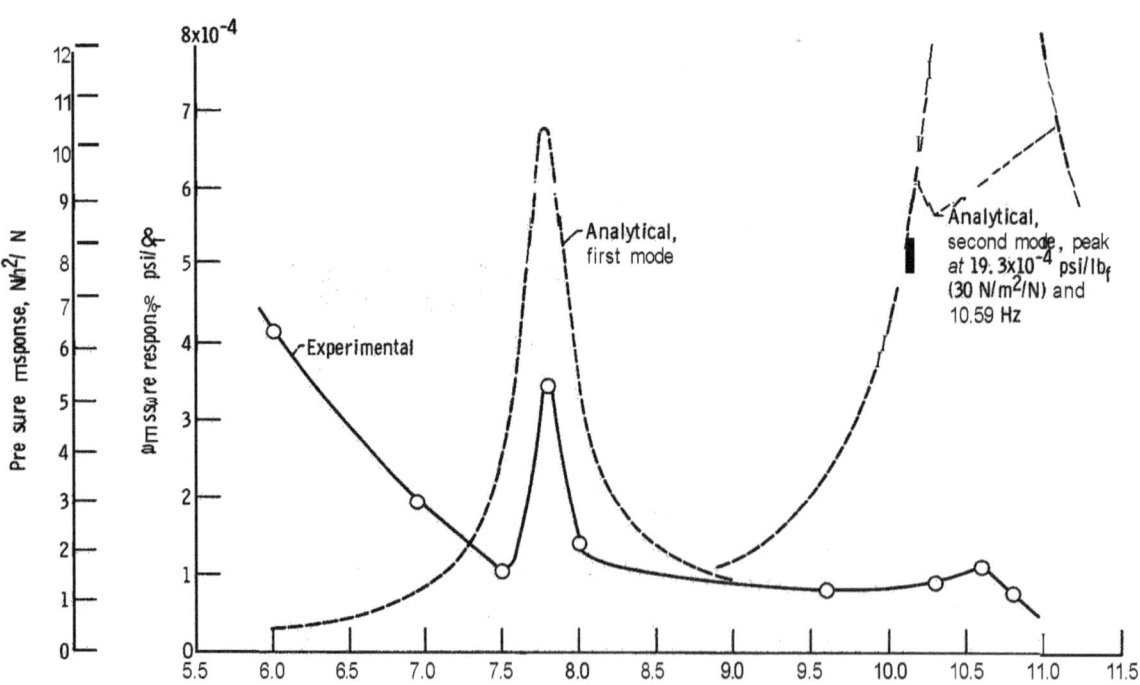

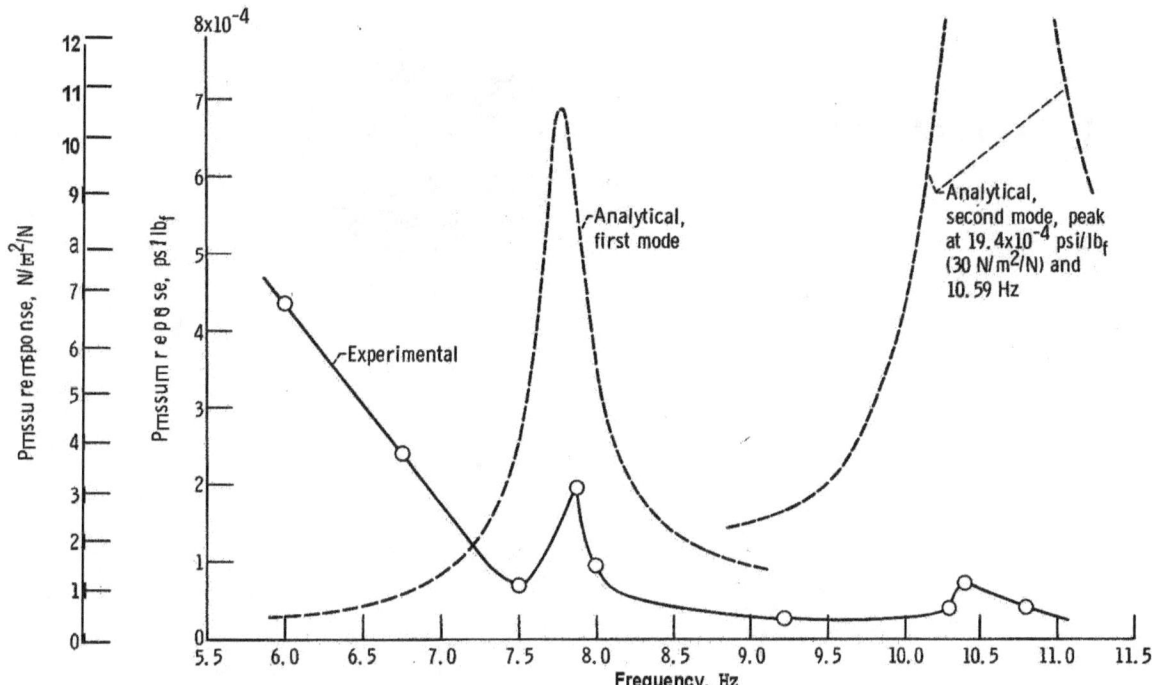

Figure 18. - Sustainer engine lox pump inlet pressure plotted against frequency. Test 10; flight time, 60 seconds.

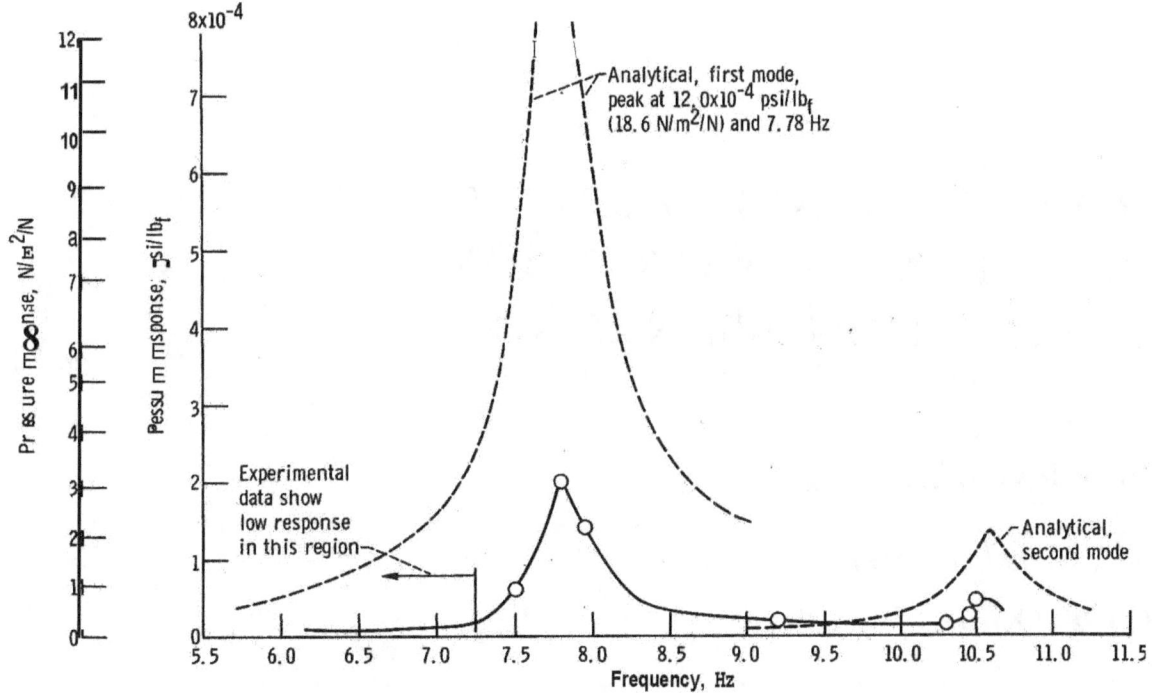

Figure 19. - Lox tank bottom pressure plotted against frequency. Test 10; flight time, 60 seconds.

NASA TECHNICAL NOTE

NASA TN D-5156

NASA TN D-5156

EXPERIMENTAL BENDING STRENGTH OF
AN ATLAS LV-3C BOOSTER BEYOND
COMPRESSIVE SKIN WRINKLING

by Robert P. Miller

Lewis Research Center
Cleveland, Ohio

NATIONAL AERONAUTICS AND SPACE ADMINISTRATION   •   WASHINGTON, D. C.   •   APRIL 1969

31

ABSTRACT

Three bending-strength tests were performed at Lewis Research Center, on a full-scale Atlas LV-3C booster to determine the bending strength available after the start of compressive wrinkling of the vehicle tank skin. Experimental data are compared with analytical predictions and show a good correlation. A description of these tests, the test results, and a method of analytically evaluating postwrinkling bending of thin-walled pressure-stabilized cylinders are presented.

32

EXPERIMENTAL BENDING STRENGTH OF AN ATLAS LV-3C BOOSTER

BEYOND COMPRESSIVE SKIN WRINKLING

by Robert P. Miller

Lewis Research Center

SUMMARY

Three bending-strength tests were conducted on a full-scale production-type Atlas booster to satisfy three general objectives: (1) to verify the existence of postwrinkling bending strength as indicated by theory and model tests, (2) to determine the bending capability of the basic tank structure, and (3) to establish a bending-strength envelope for the entire Atlas booster.

The tests were conducted at ambient conditions with the Atlas erected in a vertical position and loaded as a beam column. Data were acquired that described the deflection-against-load characteristics of the beam, stress conditions in the tank wall, and the shape of the induced skin wrinkles.

The results of the test series indicate that the behavior of the Atlas structure under the influence of loads which induce local skin wrinkling is in reasonable agreement with theory based on beam analogy assumptions and the results of model tests. The bending load capability of the middle portion of the liquid-oxygen tank was found to be 163 percent of the wrinkling onset moment under the test conditions of internal tank pressure and axial load. The wrinkles in the tank skin remained elastic until a bending moment of approximately 150 percent of the wrinkling onset moment was applied (see fig. 24), p. 36). The third test subjected the entire Atlas booster to a bending moment loading in excess of 9 million inch-pounds (1.024×10^6 m-N), which qualified areas of discontinuity on the structure to sustain postwrinkling loads.

Based on the predicted moment distribution over the Atlas-Centaur for typical flights, the test results indicate that the bending strength of the entire Atlas structure is sufficient to develop the full postwrinkling moment capability (11.2×10^6 in.-lb or 1.265×10^6 m-N) of the middle portion of the oxidizer tank (approx. station 800).

This report presents the test setup, results and discussion, and conclusions drawn from the series of three tests.

INTRODUCTION

The need for lightweight structures for aerospace applications has resulted in the use of pressure-stabilized cylindrical shells for propellant tanks of large boosters. An example of such an application is the Atlas-Centaur launch vehicle (fig. 1). During the time just preceding launch and during flight, the tank structure is subjected to severe bending from aerodynamic and inertia loadings. To obtain an efficient vehicle it is essential to have a sound understanding of bending strength and mode of failure to establish a design criterion that takes full advantage of the vehicle's strength.

In the interest of defining the full bending strength inherent in pressure-stabilized structures, considerable work has been done to develop techniques for analyzing pressurized cylinders beyond the onset of compressive skin wrinkling (see refs. 1 to 4). In support of the analysis efforts, tests have been conducted to verify the analytical findings (see refs. 5 to 9). However, most of the experimental work has been performed on small (6-in. - (0.152-m-) diam) cylinders constructed of Mylar film with no circumferential joints.

While the small-scale model tests gave good correlation with the analytical findings, there remains doubt as to how well the model test results represent the behavior of cylinders of the size, type of material, and construction found in most aerospace applications. The Atlas booster, as an example, has a diameter of 120 inches (3.042 m). The skin is made of 301 extra full-hard stainless steel varying in thickness from 0.014 inch (0.000356 m) to 0.034 inch (0.000863 m). The cylinder is formed with a series of bands, approximately 30 inches (0.761 m) wide, welded together with circumferential lap joints and doubler reinforced butt-welded vertical joints (see fig. 2).

The requirement for this investigation was based on the following objections to the use of existing knowledge:

(1) The larger diameter and radius-to-thickness ratios found in the actual boosters could result in a mode of failure not exhibited by the small-scale model.

(2) The metal skin used in actual vehicles is not as elastic as the Mylar model specimens.

(3) The lap-joint method of fabrication results in eccentricities in the tank wall which are not found in the test model specimens. These built-in eccentricities could negate the predictions established by analysis and tests based on cylinders with smooth skin.

(4) The Atlas booster differs from the small specimens by having a wiring tunnel and brackets for accessories welded to the tank wall. It was essential to observe what effect these irregularities would have on the skin wrinkling pattern.

Because of these objections to a direct application of the existing knowledge of postwrinkling strength, it was decided that a thorough understanding of the bending strength of the Atlas vehicle was beyond the state of the art and that it was therefore

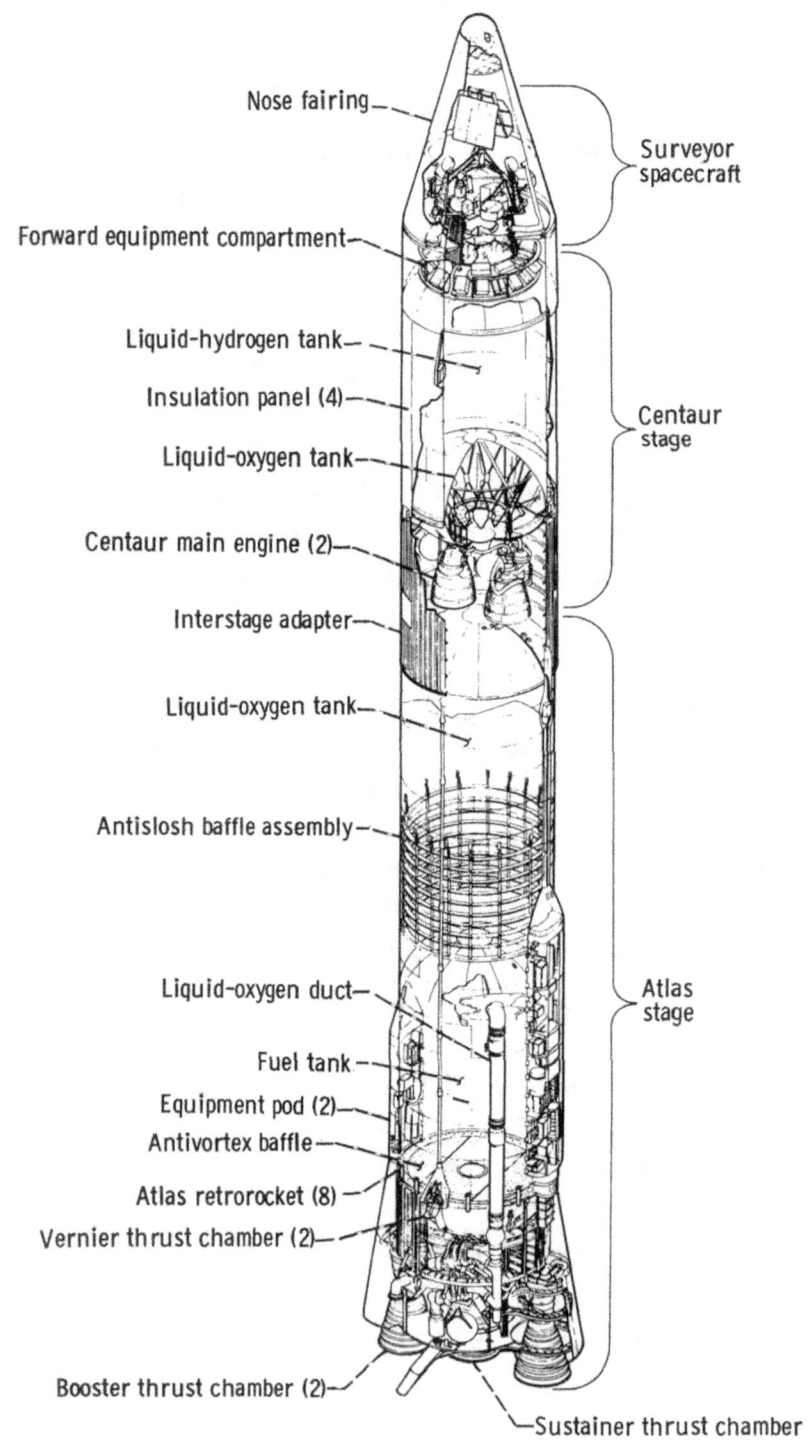

Nose fairing

Surveyor spacecraft

Forward equipment compartment

Liquid-hydrogen tank

Insulation panel (4)

Centaur stage

Liquid-oxygen tank

Centaur main engine (2)

Interstage adapter

Liquid-oxygen tank

Antislosh baffle assembly

Atlas stage

Liquid-oxygen duct

Fuel tank

Equipment pod (2)

Antivortex baffle

Atlas retrorocket (8)

Vernier thrust chamber (2)

Booster thrust chamber (2)

Sustainer thrust chamber

OC-10160-31

Figure 1. - Atlas-Centaur-Surveyor space vehicle configuration.

35

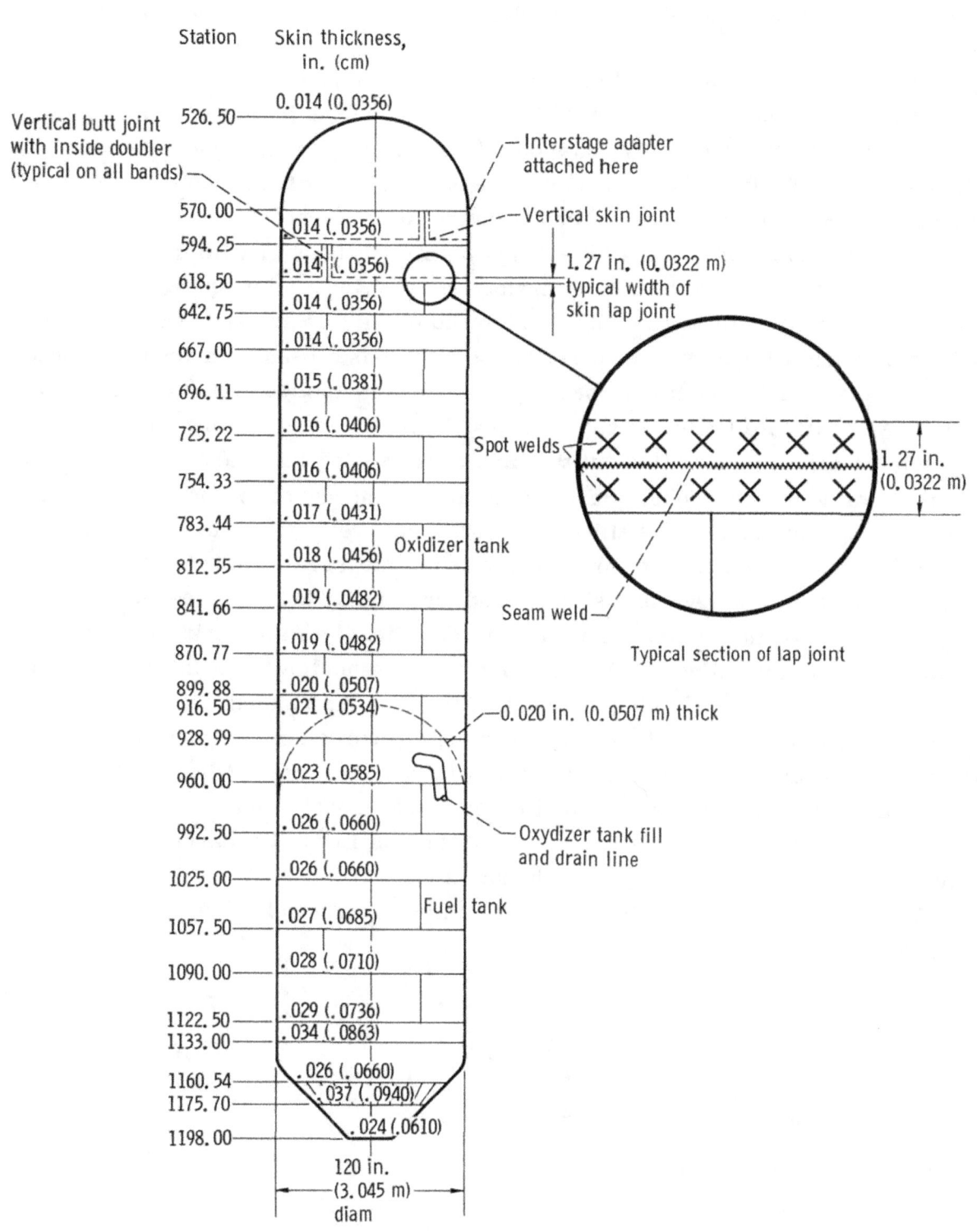

Figure 2. - Atlas test vehicle stations and propellant tank skin gages. All skins are extra full-hard 301 stainless steel.

reasonable to conduct an experimental investigation on a full-scale, production-type Atlas booster.

The investigation consisted of three tests designed to answer three specific objectives: test I was designed to verify the existence of postwrinkling bending strength in the Atlas vehicle as indicated by theory and model tests, test II was designed to determine the full bending capability of the basic tank structure, and test III was devised to establish a bending-strength envelope for the entire Atlas Booster.

The tests were carried out on a full-size Atlas LV-3C booster with the engines and their fairings removed. A flight-type interstage adapter (used to mount the Centaur second stage to the Atlas forward end) was attached to the forward end of the vehicle. The Atlas was mounted in a vertical position and loaded as a beam column with pinned ends. To simulate flight conditions, bending moment was applied simultaneously with axial load. The axial load was selected to approximate the axial loading on the Atlas during a typical Atlas-Centaur flight at the time peak bending loads are incurred (time of the maximum value of the product of angle of attack and aerodynamic pressure αQ). For all tests the propellant tanks were filled with water and pressurized to simulate flight conditions. The basic instrumentation was devised to measure strain in the tank wall, deflection of the beam, skin wrinkle shapes, and the magnitude of the applied loads.

In general, the testing followed a procedure of filling the Atlas tanks with water, establishing the specified ullage pressures, applying and holding constant the appropriate axial load, and then applying bending moment. Strain in the tank skin, deflections, and the applied loads were monitored on strip charts and recorded on magnetic tape from an analog-to-digital converter. Primary control of the loading consisted of a continuous monitoring of moment against deflection at the point of highest moment.

All of the testing described herein was performed at Lewis Research Center, Plum Brook Station, between February 1966 and August 1966.

Dr. D. J. Peery of General Dynamics/Convair was most helpful in formulating the test philosophy.

SYMBOLS

E	modulus of elasticity, psi; N/m^2	
I_{eff}	effective moment of inertia, in.^{4}; m^4	
M	external bending moment, in.-lb; m-N	
m	bending moment in skin in local wrinkle, (in.-lb)/in.; (m-N)/m	
N	maximum load, $N_t + N_c$, lb/in.; N/m	

N_c	critical wrinkling load of skin, lb/in.; N/m
N_t	maximum unit tensile load, lb/in.; N/m
N_θ	tensile hoop load in skin from internal pressure, lb/in. (N/m)
P	applied axial load, lb; N
P_a	external axial load, lb; N
P_1	axial tensile force, lb; N
p	internal pressure, psig; N/m^2 gage
Q	aerodynamic pressure
R	cylinder radius, in.; m
R_T	reaction to tower from shear loading
r	radius of curvature of centerline of elemental beam, in.; m
r_T	reaction to tower induced by deflections
r_1	radius of curvature of maximum tension fiber of elemental beam, in.; m
r_2	radius of curvature of maximum compression fiber of elemental beam, in.; m
S	applied shear load, lb; N
ΔS	elemental length of cylinder, in.; m
δS	changes in elemental length of cylinder, in.; m
t	cylinder wall (skin) thickness, in.; m
W	vehicle weight, lb; kg
w	points where vehicle weight is considered to be concentrated
y	deflection induced by applied shear or moment
α	angle of attack, deg
Δ	maximum deflection of test vehicle, in.; m
$\Delta\varphi$	angle of bend for elemental beam length, rad
ϵ_1	tensile strain, from bending and axial load only, in./in.; m/m
ϵ_2	compression strain, from bending and axial load only, in./in.; m/m
θ	half angle of unwrinkled portion of tank, deg
μ	Poisson's ratio, in./in.; m/m

$\sigma_{H,i}$ inside-surface hoop stress, psi; N/m^2

$\sigma_{H,o}$ outside-surface hoop stress, psi; N/m^2

$\sigma_{L,i}$ inside-surface longitudinal or bending stress, psi; N/m^2

$\sigma_{L,o}$ outside-surface longitudinal or bending stress, psi; N/m^2

TEST SETUP

The test vehicle for this program consisted of an Atlas LV-3C series (vehicle 116D) booster with the booster and sustainer engines removed. The Atlas LV-3C booster is a 120-inch- (3.05-m-) diameter, 563-inch- (14.3-m-) long, thin-walled cylinder with a dome-shaped bulkhead at the forward end and a conical-shaped aft end (fig. 2). An intermediate bulkhead divides the cylinder into a forward (oxidizer) tank and an aft (fuel) tank. The propellant tanks are constructed of 301 extra full-hard stainless-steel sheets varying in thickness from 0.014 inch (0.000356 m) at the forward end to 0.034 inch (0.000863 m) at the aft end. The steel sheets nominally 30 inches (0.761 m) in width are butt welded to form cylindrical bands. These bands are welded together circumferentially with lap joints to make up the total cylinder.

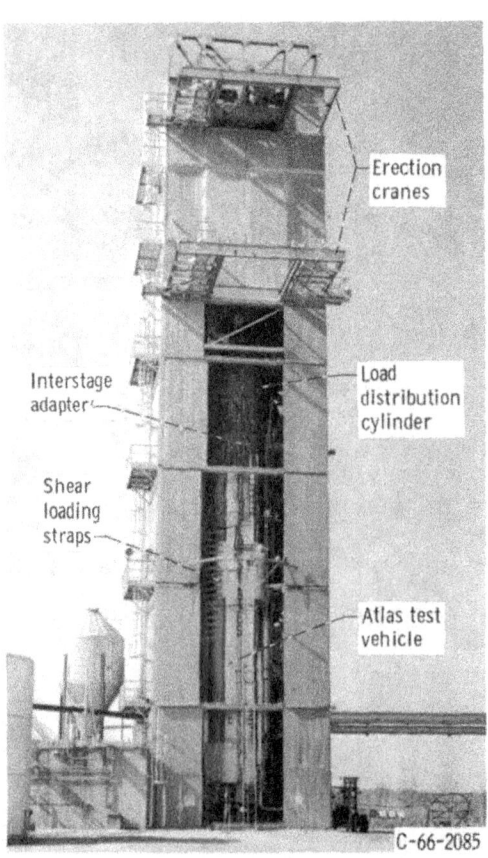

Figure 3. - Rocket Systems Dynamics Laboratory.

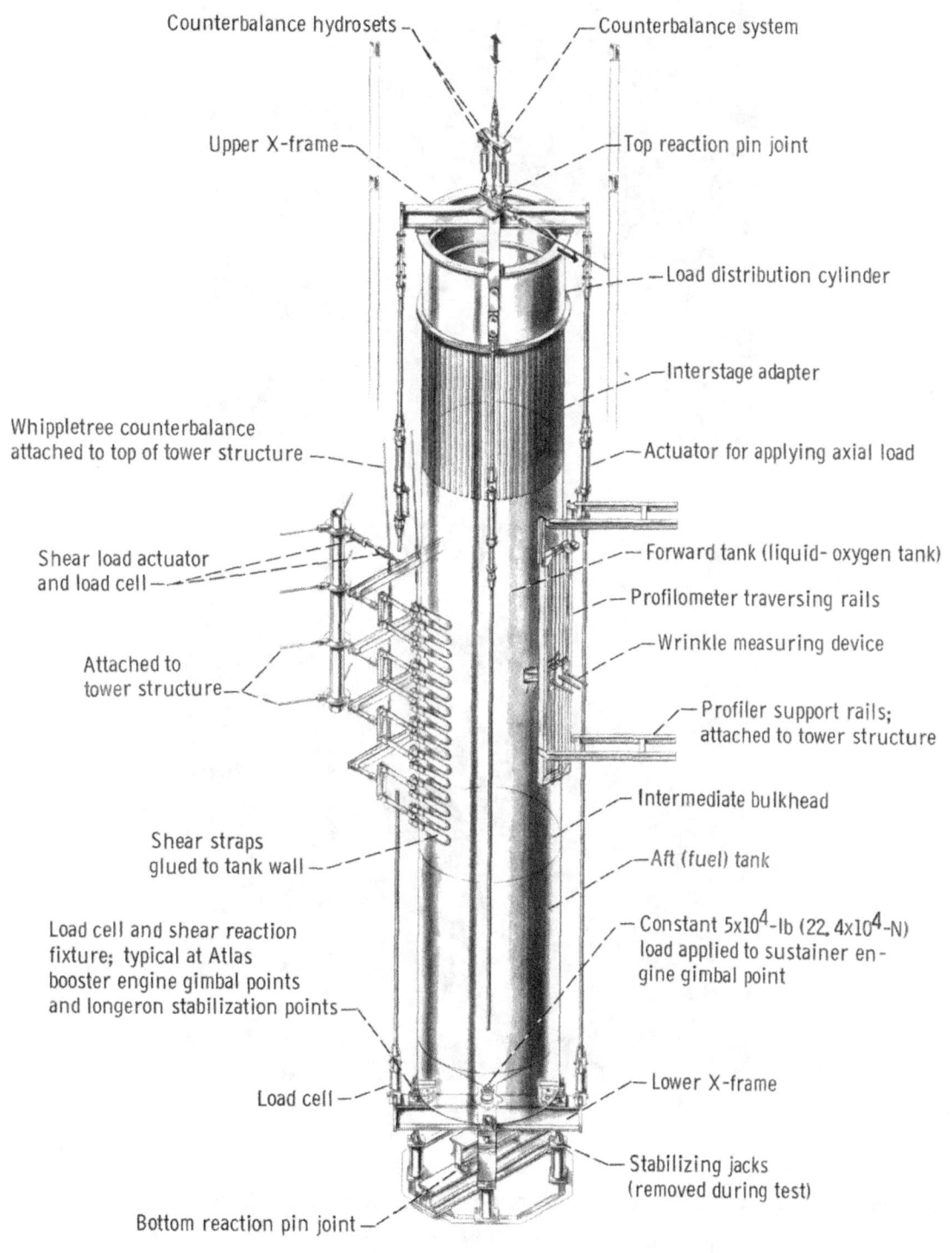

Counterbalance hydrosets

Counterbalance system

Upper X-frame

Top reaction pin joint

Load distribution cylinder

Interstage adapter

Whippletree counterbalance attached to top of tower structure

Actuator for applying axial load

Shear load actuator and load cell

Forward tank (liquid- oxygen tank)

Profilometer traversing rails

Attached to tower structure

Wrinkle measuring device

Profiler support rails; attached to tower structure

Intermediate bulkhead

Shear straps glued to tank wall

Aft (fuel) tank

Load cell and shear reaction fixture; typical at Atlas booster engine gimbal points and longeron stabilization points

Constant 5×10^4-lb (22.4×10^4-N) load applied to sustainer engine gimbal point

Load cell

Lower X-frame

Stabilizing jacks (removed during test)

Bottom reaction pin joint

CD-8438-31

Figure 4. - Postwrinkling-strength test setup for tests 1 and 2.

40

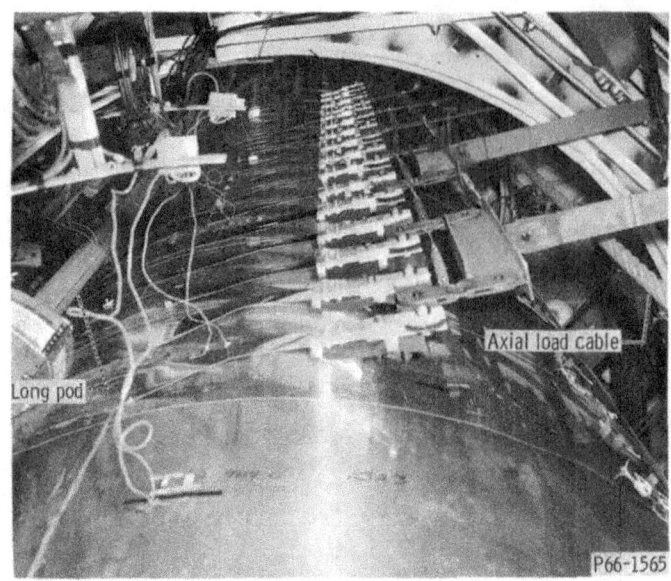

(a) View looking forward on Atlas liquid-oxygen tank.

(b) Shear-strap detail.

Figure 5. - Typical shear strap installation.

For testing, the Atlas booster was erected in the Rocket Systems Dynamics Laboratory (E tower), as shown in figure 3. The E tower is a 135-foot- (41.1-m-) high, 20-foot- (6.1-m-) square steel structure designed to accommodate an entire Atlas-Centaur vehicle.

A schematic of the test setup for the first two test conditions is shown in figure 4. The vehicle was mounted at the booster engine gimbal points and longeron support points

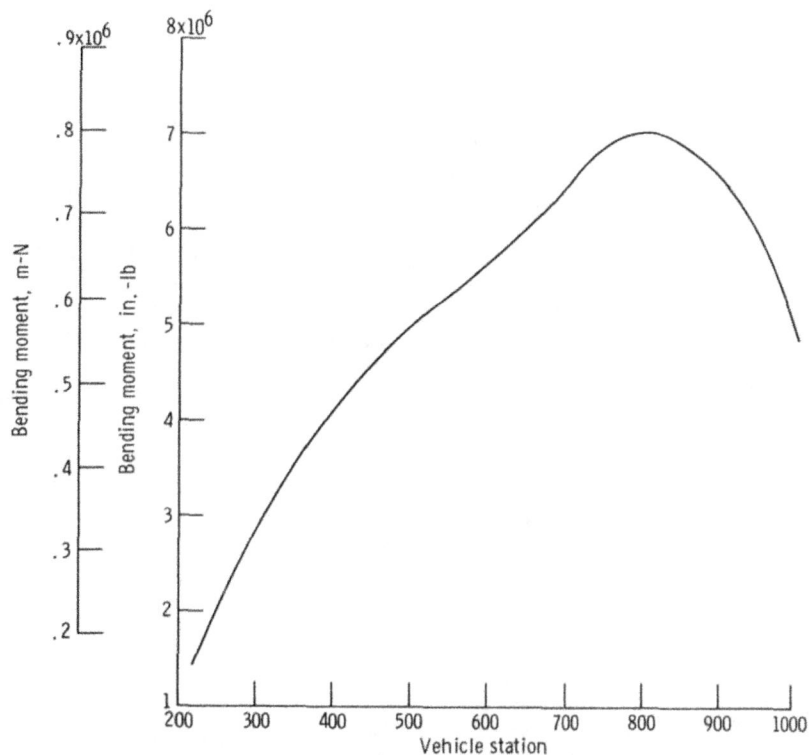

Figure 6. - Predicted bending moment as function of vehicle station at maximum value of product of angle of attack and aerodynamic pressure. Curve is based on typical March wind (see ref. 13).

to a heavy steel X-frame which in turn was attached to the floor of the tower through a pin joint. An X-frame was attached to the forward end of the vehicle through a heavy steel adapter cylinder and a flight configuration interstage adapter (used to mount the second-stage Centaur to the Atlas). The X-frame was joined to the tower structure through a pin joint that allowed free rotation and longitudinal motion. Axial compression load was applied through hydraulic actuators located in series with four cables strung the length of the vehicle and attached to the X-frames at each end of the test vehicle.

In the first two tests of the series, bending was achieved with shear loading applied to the Atlas. The shear load was induced through 16 pairs of steel straps glued along the neutral axis on each side of the oxidizer tank with an air-curing silicone rubber cement (see fig. 5). Loading of the straps was accomplished with four hydraulic actuators whippletree-connected to the glued-on straps. Hydraulic pressure to the four actuators was controlled through a proportioning device that allowed shaping the bending moment curve to approximate that predicted in flight (fig. 6).

To provide high bending moment over the entire length of the vehicle, as required for the third test, the shear straps were removed and bending was accomplished by differential loading of two cables strung the length of the vehicle. To distribute the con-

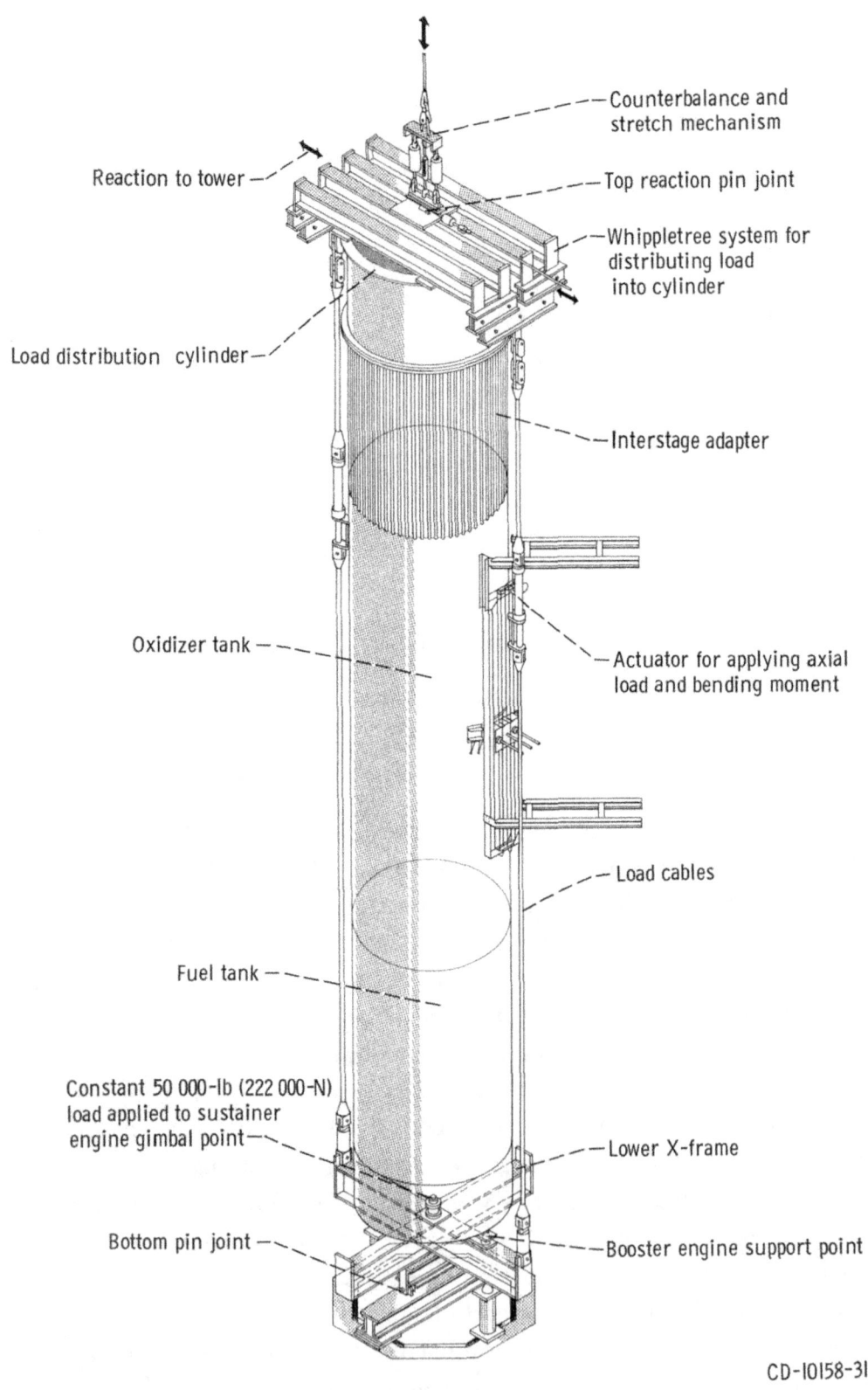

Counterbalance and stretch mechanism

Reaction to tower

Top reaction pin joint

Whippletree system for distributing load into cylinder

Load distribution cylinder

Interstage adapter

Oxidizer tank

Actuator for applying axial load and bending moment

Load cables

Fuel tank

Constant 50 000-lb (222 000-N) load applied to sustainer engine gimbal point

Lower X-frame

Bottom pin joint

Booster engine support point

CD-10158-31

Figure 7. - Postwrinkling-strength test setup for test 3.

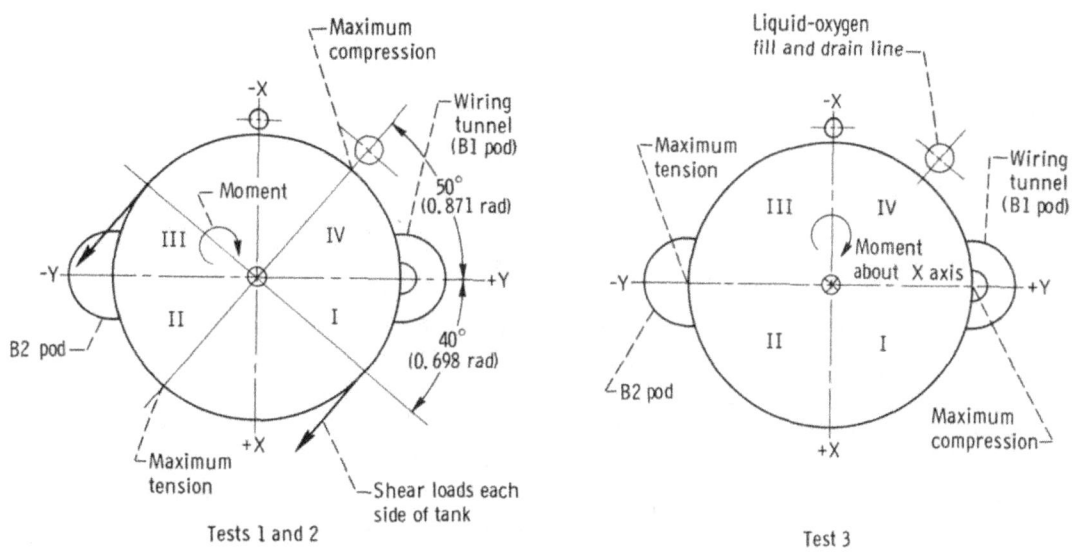

Figure 8. - Bending axis orientation.

centrated load from the two cables evenly into the Atlas tank, the top X-frame was removed and replaced with a system of beams placed across the top of the load distribution cylinder. The loading cables were attached to the system of beams through a whippletree arrangement to distribute the cable load into the load distribution cylinder. (A schematic of the test 3 configuration is presented in fig. 7).

The loading systems for the three tests oriented the loads to the vehicle, as shown in figure 8.

INSTRUMENTATION

The Atlas was instrumented to measure lateral and longitudinal deflection with string-type displacement transducers located as shown in figure 9. The transducers for measuring the bending deflection of the tank were attached to the tension side of the tank wall. These measurements were made relative to the tower structure. Transit sightings of targets attached to the tower were made to ascertain that the tower provided a stable reference throughout testing. The electrical transducer deflection measurements were obtained with an accuracy of ±0.05 inch (±0.00127 m).

In order to accurately define the mode of wrinkling in the tank wall, a remotely operated device (profilometer) was designed to measure the profile shape of the wrinkles (fig. 10). The wrinkle shapes were recorded on an X-Y plotter. The profilometer

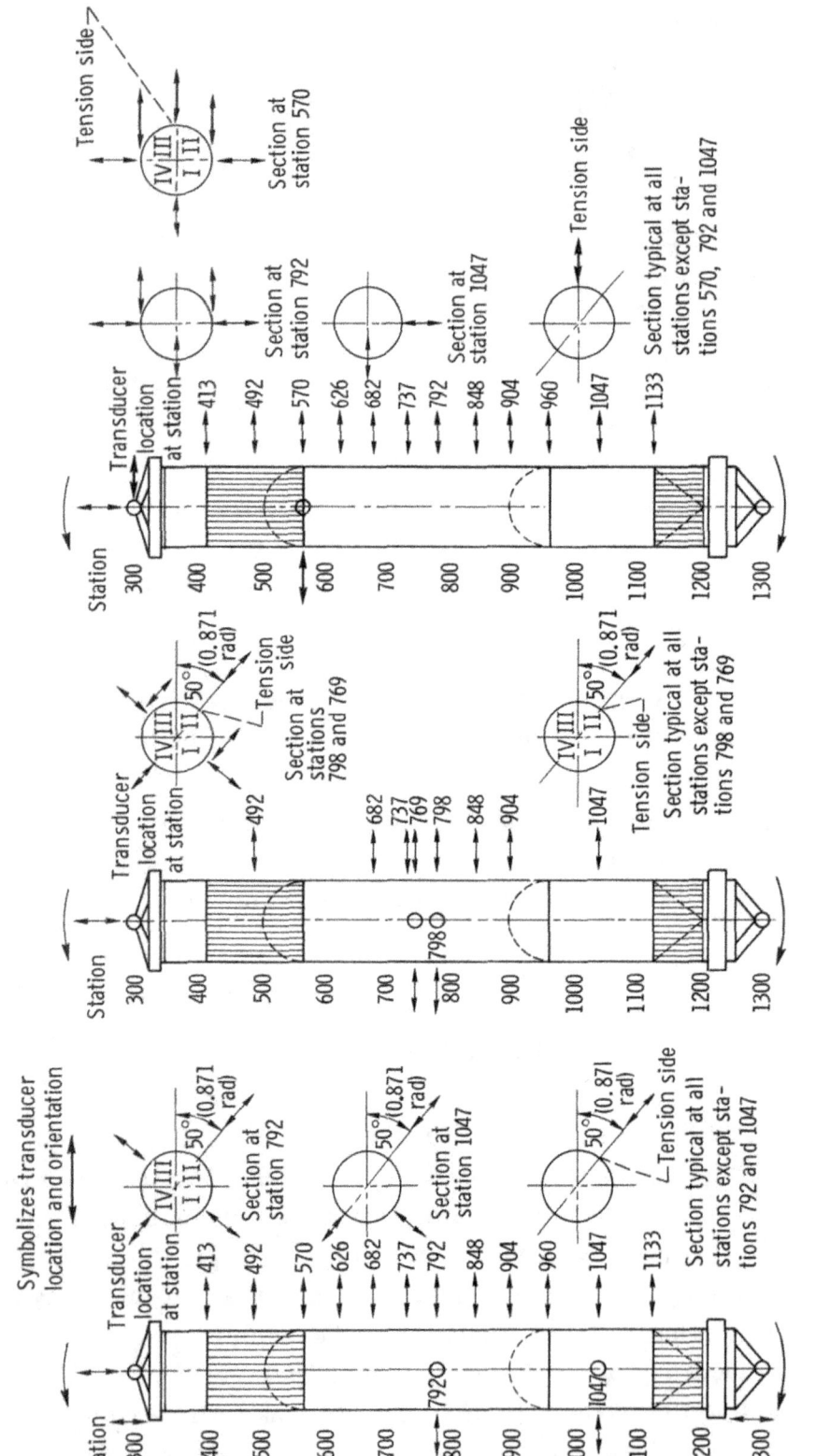

Figure 9. – Displacement transducer locations.

Atlas 116D – Part 1

45

(a) Installation.

(b) Detail.

Figure 10. – Wrinkle-measuring-device (profilometer) installation.

46

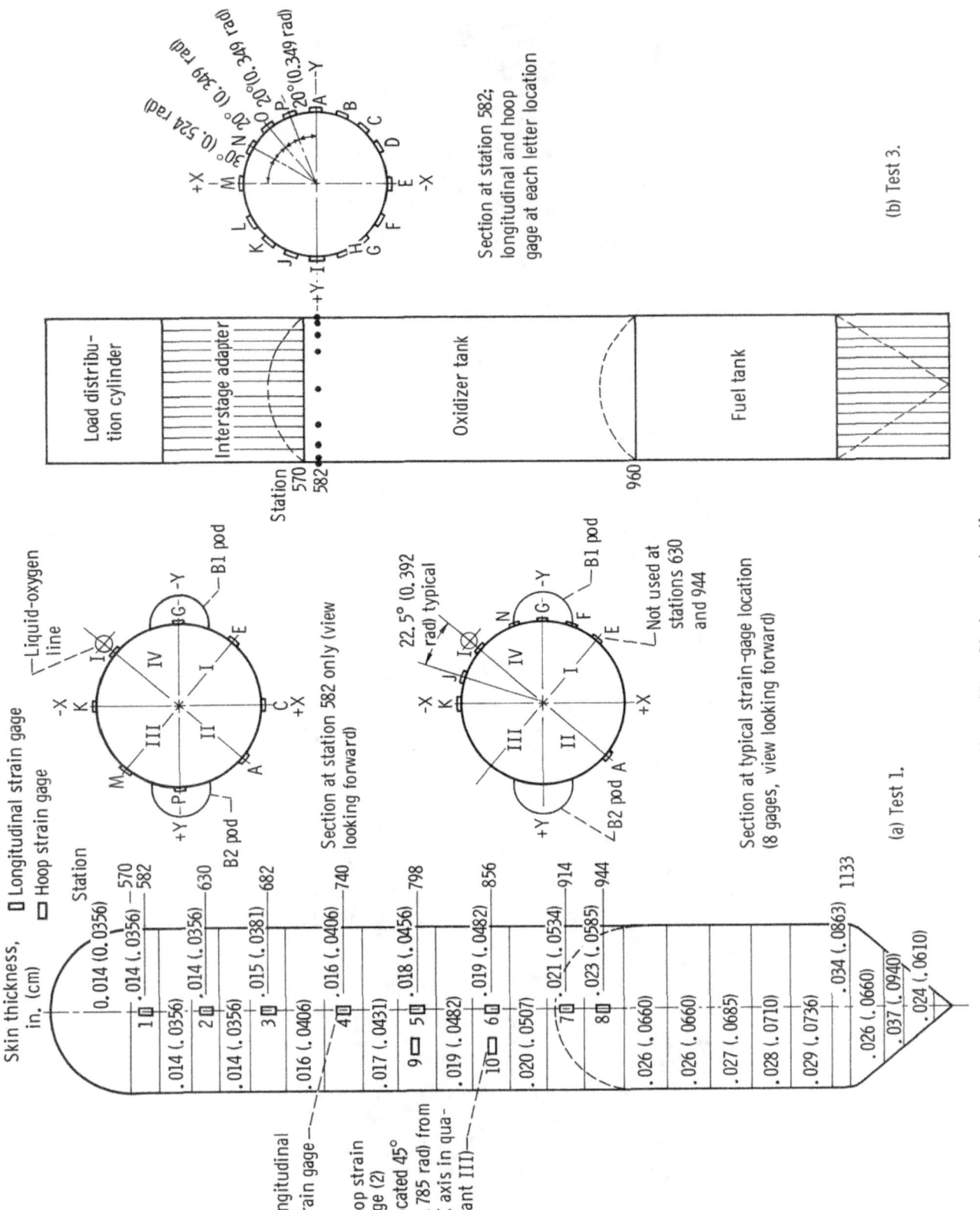

Figure 11. - Strain-gage locations. - Part 1

(a) Test 1.

(b) Test 3.

47

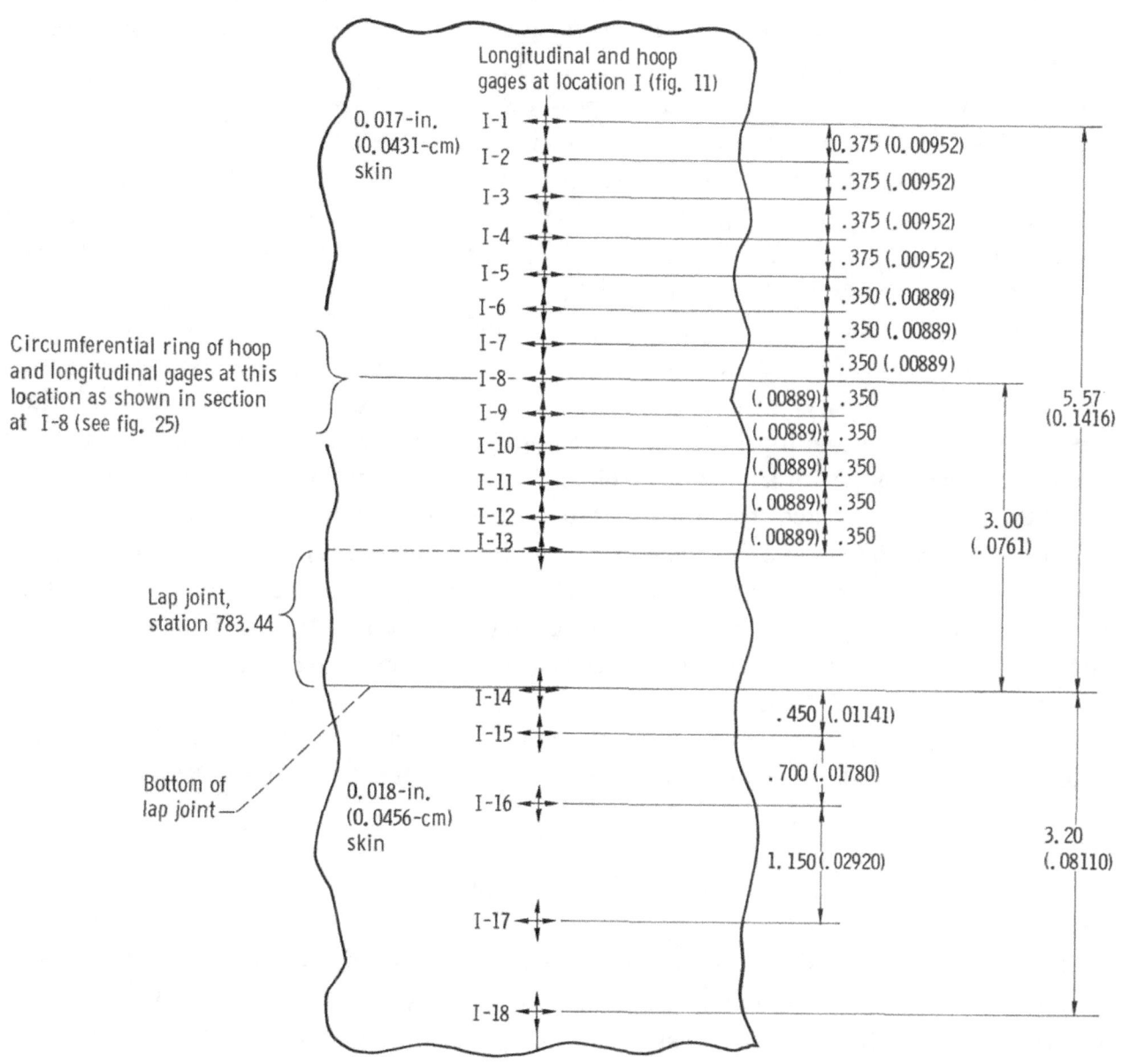

Figure 12. - Strain-gage locations for test 2. Dimensions are in inches (m).

48

mechanism was capable of traversing a 15-foot (4.56-m) length of the oxidizer tank in the area of maximum compression. The profilometer consisted of three displacement transducers mounted on a carriage that traversed the length of the oxidizer tank by means of guide rails and a chain-drive arrangement. The transducers sensed displacements through spring-loaded arms that maintained contact with the tank surface while the carriage traversed. The instrument was capable of being controlled to engage the tank wall while collecting data and to retract while loading was taking place. Profilometer traces of the tank wall were made after each loading condition was established.

For tests 1 and 2, strain gages were located on the outside surface of the tank wall around the circumference at various stations (see fig. 11). These gages were oriented to indicate both longitudinal and hoop strain.

For test 2, additional strain gages were placed over a point of high wrinkling as indicated by the profilometer in test 1. The gages were mounted on the outside surface of the tank in the area of the lap joint at station 783, as shown in figure 12. These gages gave an indication of the local hoop and bending strain in the wrinkles. All strain measurements were made with an accuracy of ±70 microinches per inch (±1.78 μm/m).

All testing was controlled remotely from a central control building approximately 1/4 mile (402 m) from the test tower to eliminate the hazard from a potential tank rupture. Remotely operated cameras recorded pictures of the test vehicle at selected increments of loading. Continuous surveillance of the Atlas was maintained from the control building by means of television cameras and through high-power telescopes located 250 feet (76 m) from the test tower.

The strain-gage, displacement-transducer, and load-cell outputs were recorded on magnetic tape through an analog-to-digital converter at the control building.

TEST PROCEDURE

The test procedures for tests 1 and 2 were essentially the same, except that the bending in test 2 was taken to the full moment capability of the oxidizer tank. The Atlas was first filled with water and pressurized, the fuel tank to 58.5 psig (403 000 N/m^2 gage) and the oxidizer tank to 24 psig (165 000 N/m^2 gage). An axial compressive load of 75 800 pounds (336 000 N) was applied and held constant throughout each of the first two tests. The bending was accomplished through loading of the shear straps. Bending moment was applied in one-million-inch-pound (0.113×10^6-m-N) increments with data being recorded at each stabilized load increment. The applied bending moment was defined by the shear strap loads acting through the geometry of the test setup. A load cell, mounted in conjunction with the top pin connection, gave verification of the applied shear through its reaction. The shear loads were measured with an accuracy of 150 pounds (665 N) by load cells located in series between each hydraulic actuator and its corresponding set of straps.

During test 1 the moment load was increased and then dropped back to a base level $(4 \times 10^6$ in. -lb, or 0.451×10^6 m -N) that was well below the wrinkling onset point, in order to investigate the repeatability of the loadings and deflection. For test 2 the load was increased in increments to the maximum level $(10.2 \times 10^6$ in. -lb, or 1.15×10^6 m-N) without returning to the base value.

A continuous plot of deflection against bending moment at station 798 in the oxidizer tank was maintained during the test. These values were compared with predicted values as a primary indication of bending capability remaining before the next increment of load was applied. At each load, data from the profilometer were studied as a check of the severity of the skin wrinkles. Key strain gages were also monitored during the test to measure distribution of load into the vehicle.

The procedure for test 3 followed that for the first two tests, with the following exceptions. The oxidizer tank pressure was set at the flight pressure of 28.5 psig (196 000 N/m^2 gage), and the axial load was maintained at the simulated flight axial load of 126 000 pounds (559 000 N). The shear straps were removed, and a constant bending moment was induced over the entire length of the tank by differential loading of two longitudinal cables in the Y, Z plane. In this manner a couple was induced that essentially applied a constant moment over the length of the Atlas and adapters. The same procedure for monitoring the test progress was followed as for tests 1 and 2.

RESULTS AND DISCUSSION

The test program consisted of three tests. Each test essentially fulfilled one of the three basic objectives of the program.

The first test (test 1) of the series was designed to verify the postwrinkling behavior of the full-scale Atlas vehicle as indicated by analysis and previous model testing (refs. 2 and 3).

The maximum bending load applied in test 1 is· shown in figure 13, along with the measured and predicted deflection of the vehicle for the peak load condition. (The method of obtaining the analytical deflections is presented in appendix A of this report.) The maximum moment applied was 10.2×10^6 inch-pounds $(1.15 \times 10^6$ m-N) at station 810. This moment is approximately 50 percent higher than the predicted wrinkling onset load designated by the dashed line in figure 13. The comparison of predicted and measured deflection indicates that the beam deflection of the wrinkled Atlas vehicle can be predicted with reasonable accuracy.

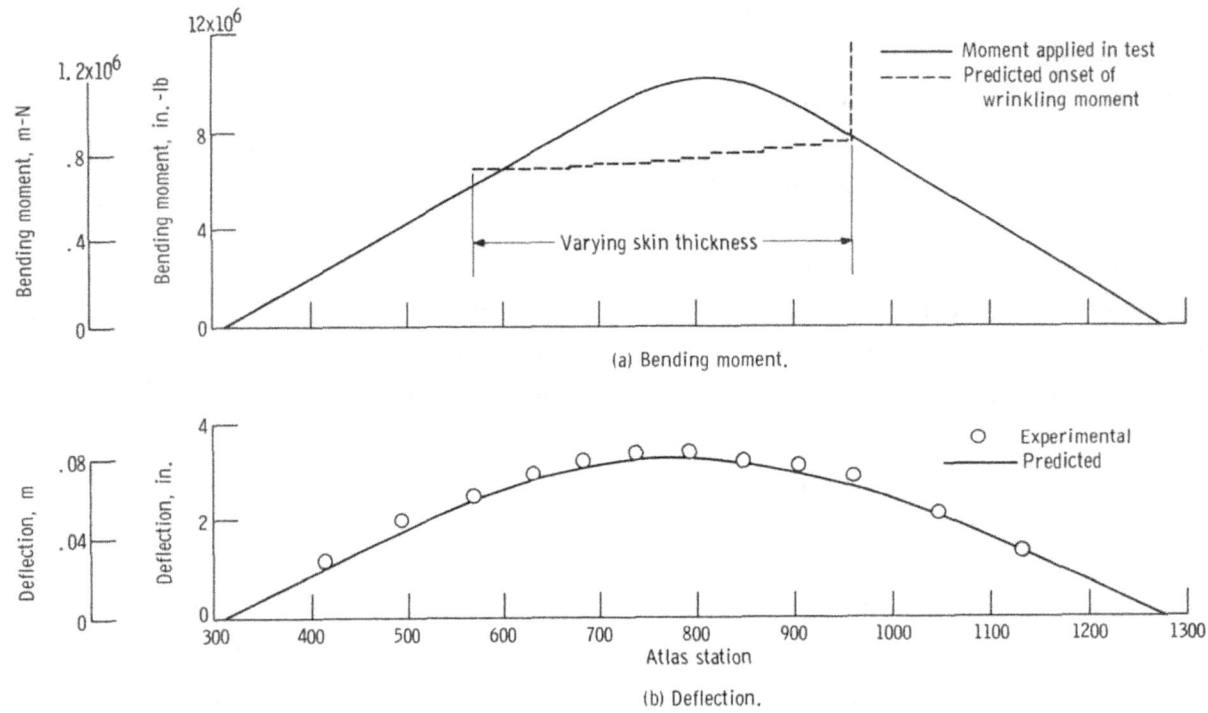

(a) Bending moment.

(b) Deflection.

Figure 13. - Bending moment and deflection as function of Atlas station - test 1. Test conditions: axial load, 75 800 pounds (336 000 N); oxidizer tank pressure, 24.0 psig (165 000 N/m²); fuel tank pressure, 58.5 psig (402 500 N/m²); oxidizer tank filled with water to station 540; fuel tank filled with water to station 925.

As the moment was increased beyond the onset of skin wrinkling the buckles appeared as damped sinusoidal waves propagating from the lap joint areas and progressing toward the middle portion of the tank skins (see fig. 14). The profile of the tank surface at the extreme compression fiber is shown in figure 15 under the influence of several test 1 loading conditions.

The stress distribution around the tank circumference as measured with strain gages is compared with predicted distribution in figure 16. The analytical load or stress distribution around the tank was obtained by using the equations presented in appendix B of this report. The stress or load distribution around the circumference of the tank is in reasonable agreement with theory. A plot of deflection against bending moment, both predicted and experimental, at the station of maximum moment is shown in figure 17. The deflection data points are numbered to indicate the sequence in which they were acquired and to illustrate the effect of reducing the moment to the base value before proceeding to the next higher load increment. The linearity in repeating the wrinkling loads was reasonably good with the deflection repeating at the base load within 0.2 inch (0.00507 m). It is believed that at least half of this value was the result of taking up test fixture slack and of self-alinement.

51

(a) Axial load only.

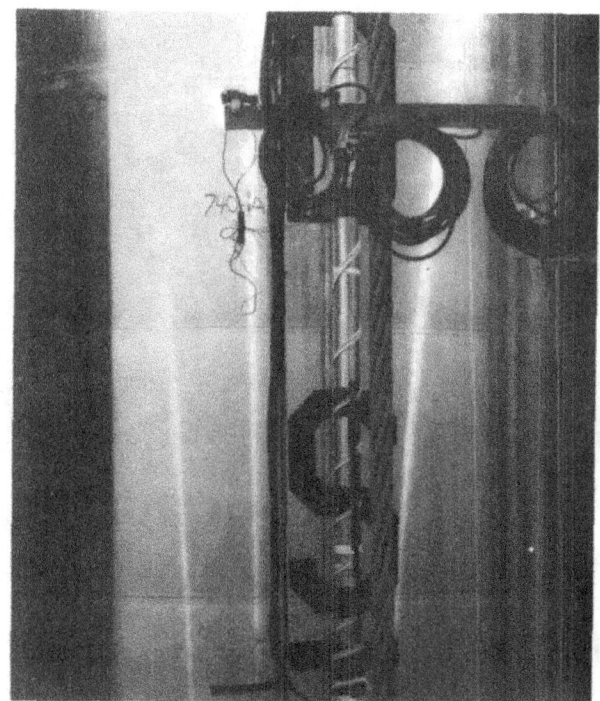

(b) Bending moment, 4.23×10^6 inch-pounds (0.476×10^6 m-N).

(c) Bending moment, 7.42×10^6 inch-pounds (0.837×10^6 m-N).

(d) Bending moment, 7.96×10^6 inch-pounds (0.900×10^6 m-N).

Figure 14. - Wrinkle patterns for test 1. Axial load, 75 800 pounds (336 000 N).

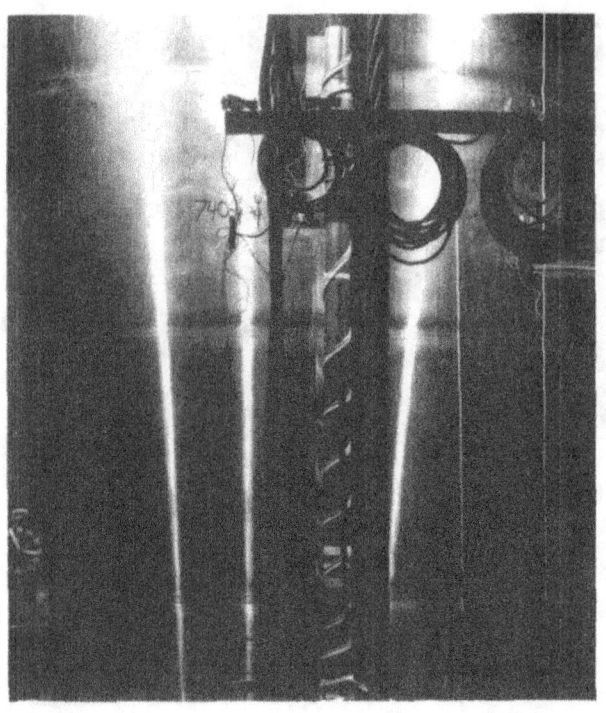

(e) Bending moment, 8.51×10^6 inch-pounds $(0.961 \times 10^6$ m-N).

(f) Bending moment, 9.06×10^6 inch-pounds $(1.023 \times 10^6$ m-N).

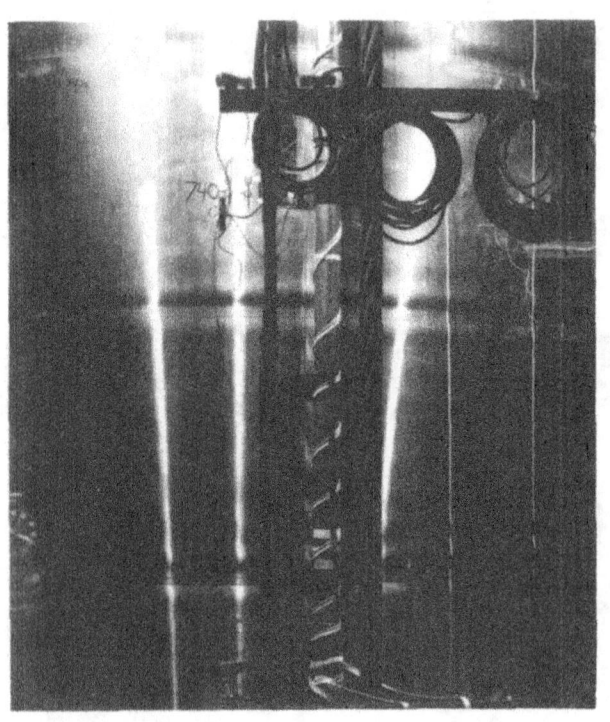

(g) Bending moment, 9.62×10^6 inch-pounds $(1.089 \times 10^6$ m-N).

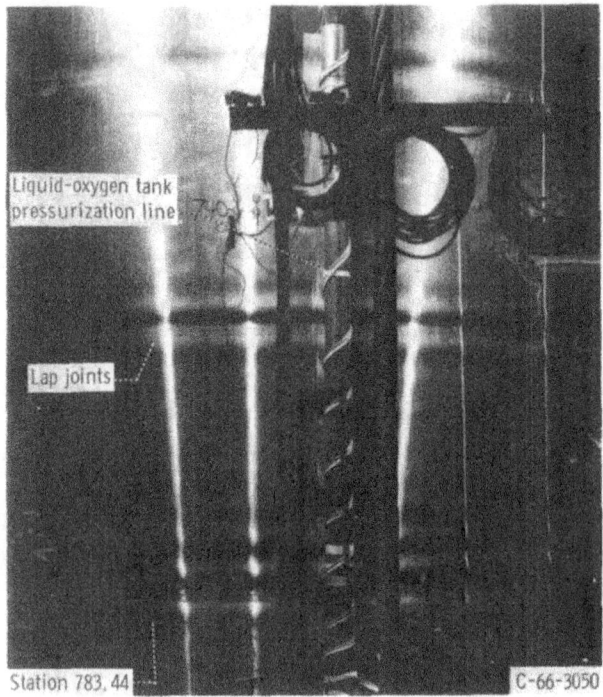

(h) Bending moment, 10.22×10^6 inch-pounds $(1.158 \times 10^6$ m-N).

Figure 14. - Concluded.

(page missing in source file)

(page missing in source file)

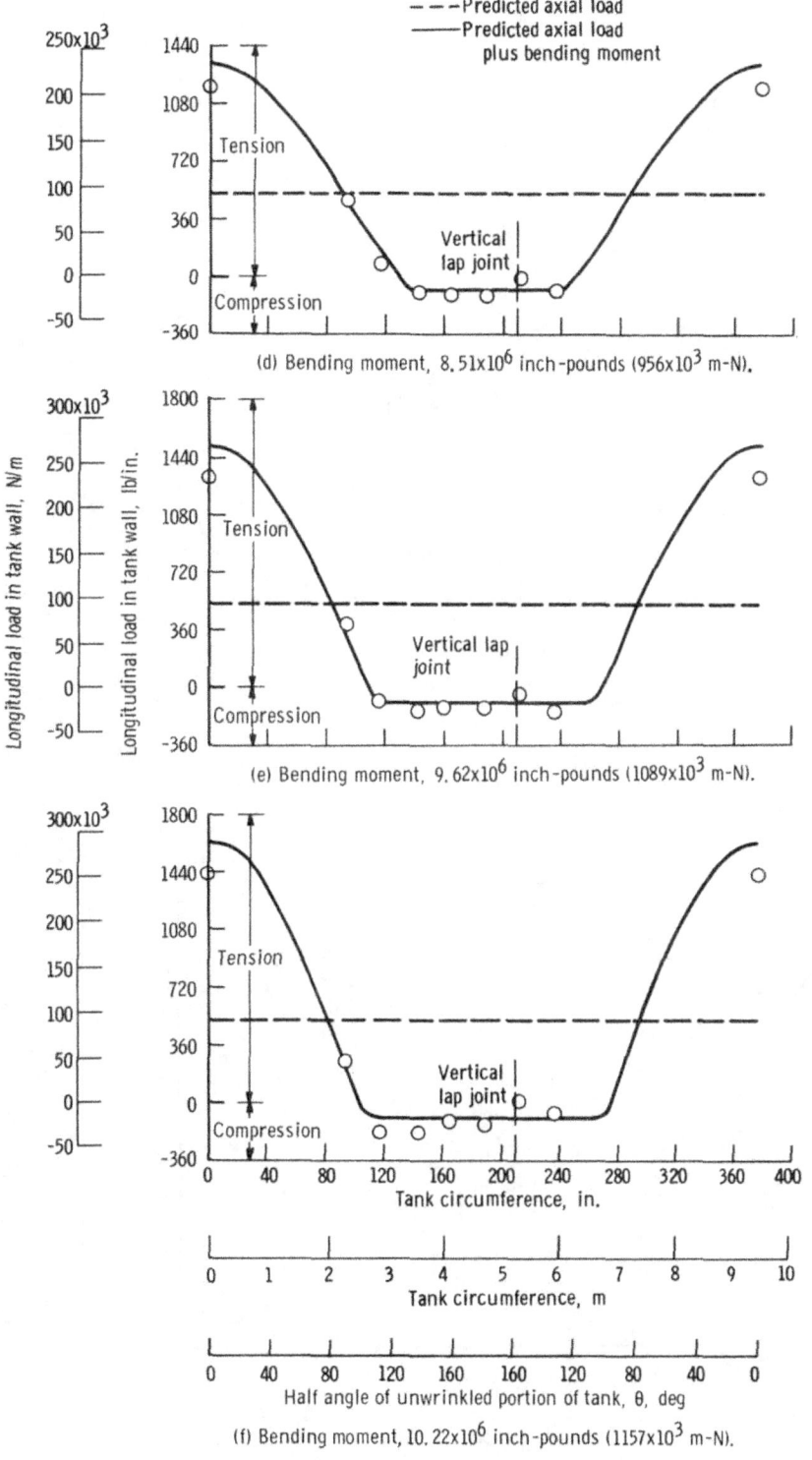

(d) Bending moment, 8.51×10^6 inch-pounds (956×10^3 m-N).

(e) Bending moment, 9.62×10^6 inch-pounds (1089×10^3 m-N).

(f) Bending moment, 10.22×10^6 inch-pounds (1157×10^3 m-N).

Figure 16. - Concluded.

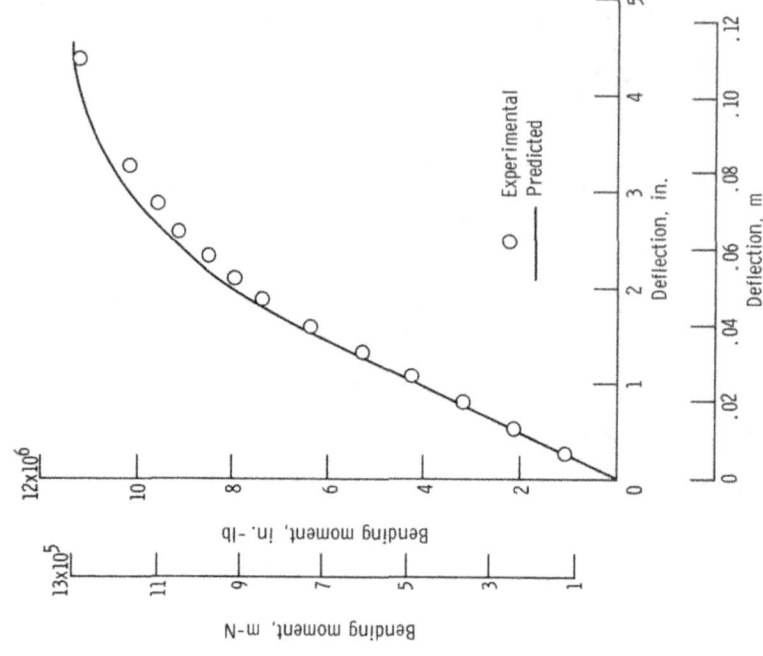

Figure 18. - Comparison of theoretical deflection with data at station 798 (approx. middle of oxidizer tank). Test 1; repeated loading; axial load, 75 800 pounds (336 000 N).

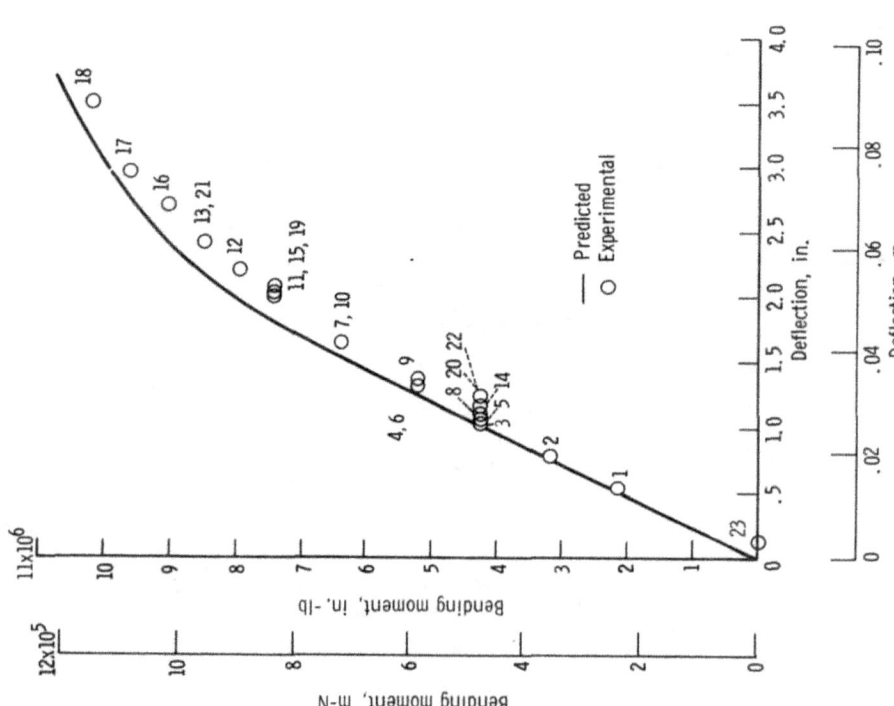

Figure 17. - Comparison of theoretical deflection with data at station 792. Test 1; repeated loading; axial load, 75 800 pounds (336 000 N). (Numbers accompanying data points indicate sequence of load application.)

57

The results of test 1 indicate that the behavior of the Atlas structure under the influence of postwrinkling bending loads follows the analysis of reference 1 and the findings of model tests (refs. 2, 3, 5, and 10) as to primary bending stability, mode and extent of skin wrinkling, and beam deflection. (More details of test 1 are presented in ref. 10.)

The specific intent of test 2 was to determine the maximum moment capability of the Atlas oxidizer tank under simulated flight conditions of axial load and tank pressure and to confirm the findings of test 1. In addition, strain-gage instrumentation was applied to study the local stress condition of the wrinkled skin (see fig. 12).

In test 2, the vehicle was loaded in the same manner as test 1. The loading was continued until the secondary moment induced by the vehicle deflection was sufficient to make the beam unstable without further application of shear load (see fig. 18). The maximum moment loading on the vehicle in test 2 is shown in figure 19(a), and the predicted and measured deflection of the Atlas under the maximum load condition in figure 19(b). The peak moment of 11.2×10^6 inch-pounds $(1.265 \times 10^6$ m-N) is 63 percent above the predicted onset of wrinkling moment, shown as a dashed line in figure 19(a), and 17 percent below the theoretical ultimate moment of 13.5×10^6 inch-pounds

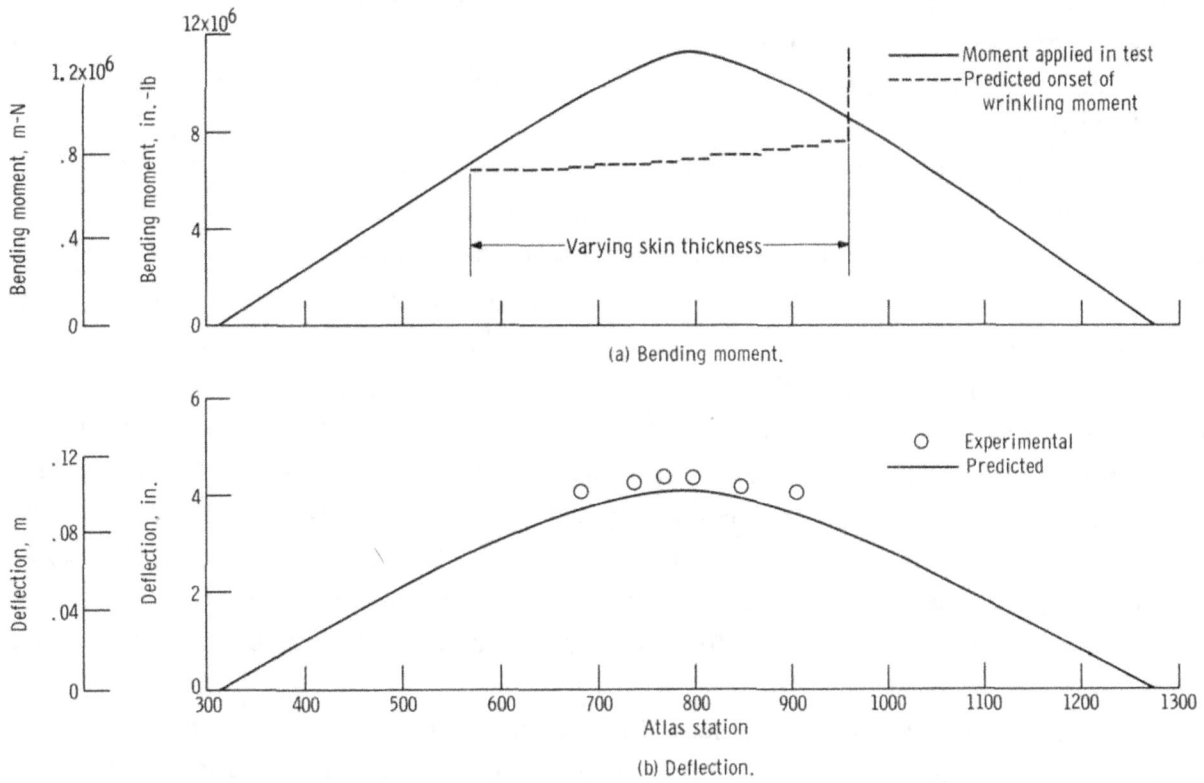

Figure 19. - Bending moment and deflection as function of Atlas station - test 2. Test conditions: axial load, 75 800 pounds (336 000 N); oxidizer tank pressure, 24.0 psig (165 000 N/m^2); fuel tank pressure, 58.5 psig (402 500 N/m^2); oxidizer tank filled with water to station 540; fuel tank filled with water to station 925.

(1.525×10^6 m-N) (ref. 1), which would have produced wrinkles around the entire circumference of the tank. The 11.2×10^6-inch-pound (1.265×10^6-m-N) moment produces wrinkling around approximately 202° of the tank circumference. The peak bending moment was reached, however, before the wrinkles enveloped the entire tank circumference. This occurred when the shallow damped sinusoidal wrinkle pattern abruptly shifted to a few deep wrinkles concentrated at one lap joint (see fig. 20). This abrupt shift of wrinkle pattern was also observed in the testing of reference 5.

Test 2 was terminated when static equilibrium could no longer be maintained, and all loads were immediately removed to prevent destruction of the test vehicle. Typical wrinkle patterns observed during test 2 are shown in both figures 20 and 21. In figure 21, skin wrinkling can be observed under the wiring tunnel. The presence of the tunnel had no apparent effect on the wrinkling pattern nor was any damage to the tunnel and its attachments observed.

The stress levels measured at station 783 are presented in figure 22. In this figure the hoop and longitudinal stresses at the extreme compression fiber are compared with the buckle profile as established with the profilometer in the instrumented area. The figure presents a progression of stress level as the test loads were applied. The stress values were derived from the biaxial strain readings using modulus of elasticity values of 25.5×10^6 and 29.3×10^6 psi 175×10^9 and 202×10^9 N/m^2), respectively, for the hoop and longitudinal directions.

The bending moment capability of the test vehicle is illustrated in figure 18. Figure 23(a) compares the principal stresses on the outside surface of the tank with applied moment. The stress values were those obtained at the most highly stressed point in the wrinkled area at station 783. The peak stresses occurred in the first outward wrinkle above the lap joint at station 783. All stresses were obtained from strain-gage instrumentation on the outside surface of the tank skin. As bending moment was applied to the vehicle, the hoop stress remained relatively constant until the onset of skin wrinkling. After wrinkling the hoop stress increased linearly with moment. As the vehicle was bent, the longitudinal skin tension induced by the tank pressure was relieved on the compression side until the skin began to wrinkle. As wrinkling began, the longitudinally oriented gages indicated increasing tension, primarily from local bending of the skin in the area instrumented. The longitudinal stress assumed a linear relation with moment after the onset of wrinkling.

Figure 23(b) presents the corresponding stress conditions on the inside surface of the tank wall. These values were calculated by using the experimental strain data collected on the outside surface. The calculations were made as follows:

(a) Bending moment, 6.35x10^6 inch-pounds (716x10^3 m-N).

(b) Bending moment, 7.4x10^6 inch-pounds (835x10^3 m-N).

(c) Bending moment, 7.95x10^6 inch-pounds (898x10^3 m-N).

(d) Bending moment, 8.5x10^6 inch-pounds (960x10^3 m-N).

(e) Bending moment, 9.05x10^6 inch-pounds (1021x10^3 m-N).

(f) Bending moment, 9.6x10^6 inch-pounds (1085x10^3 m-N).

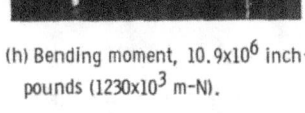

(g) Bending moment, 10.15x10^6 inch-pounds (1146x10^3 m-N).

(h) Bending moment, 10.9x10^6 inch-pounds (1230x10^3 m-N).

(i) Bending moment, 11.04x10^6 inch-pounds (1250x10^3 m-N).

Figure 20. - Wrinkle patterns - test 2.

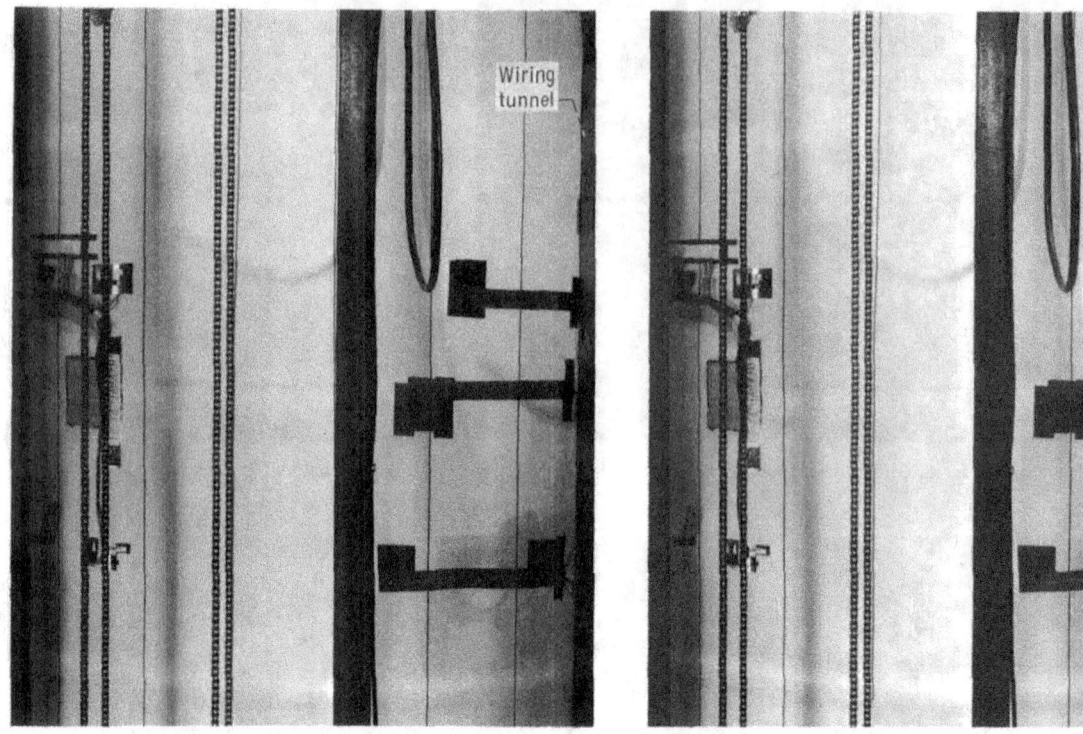

(a) Bending moment, 6.35×10^6 inch-pounds (716×10^3 m-N).

(b) Bending moment, 7.4×10^6 inch-pounds (835×10^3 m-N).

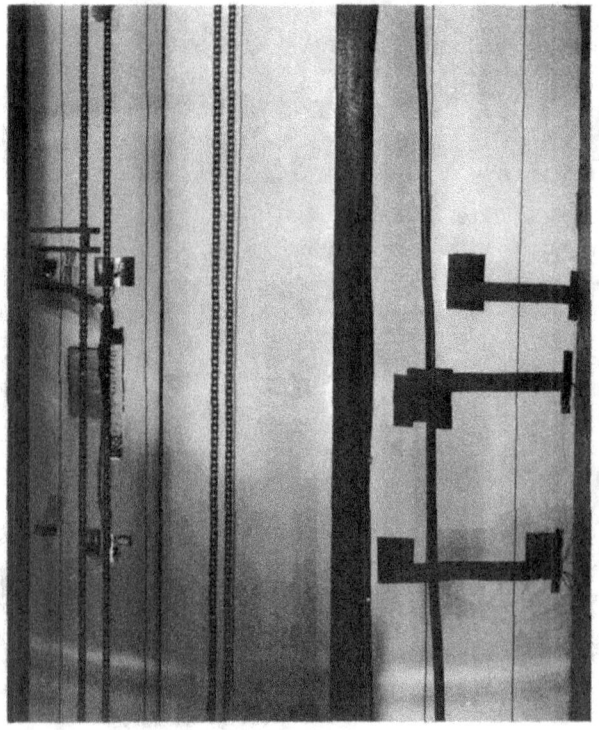
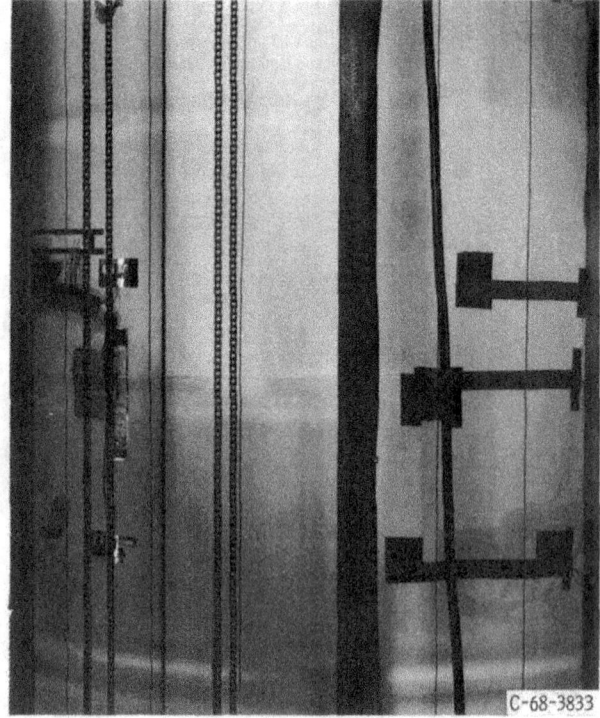

(c) Bending moment, 7.95×10^6 inch-pounds (898×10^3 m-N).

(d) Bending moment, 8.5×10^6 inch-pounds (960×10^3 m-N).

Figure 21. - Closeup of wrinkle patterns - test 2.

61

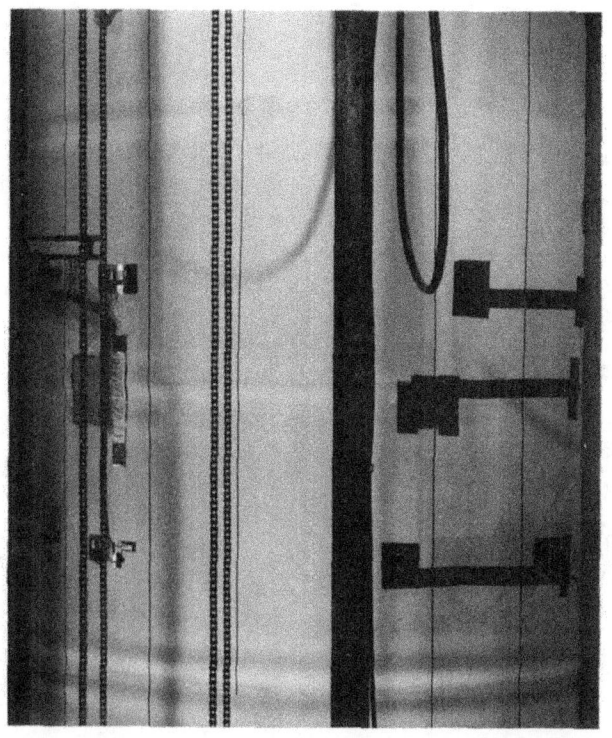

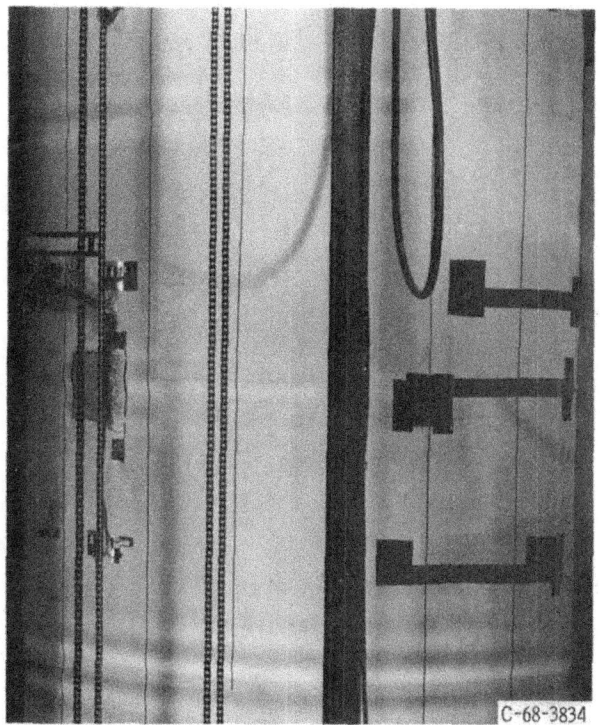

(e) Bending moment, 9.05x10⁶ inch-pounds (1021x10³ m-N).

(f) Bending moment, 9.6x10⁶ inch-pounds (1085x10³ m-N).

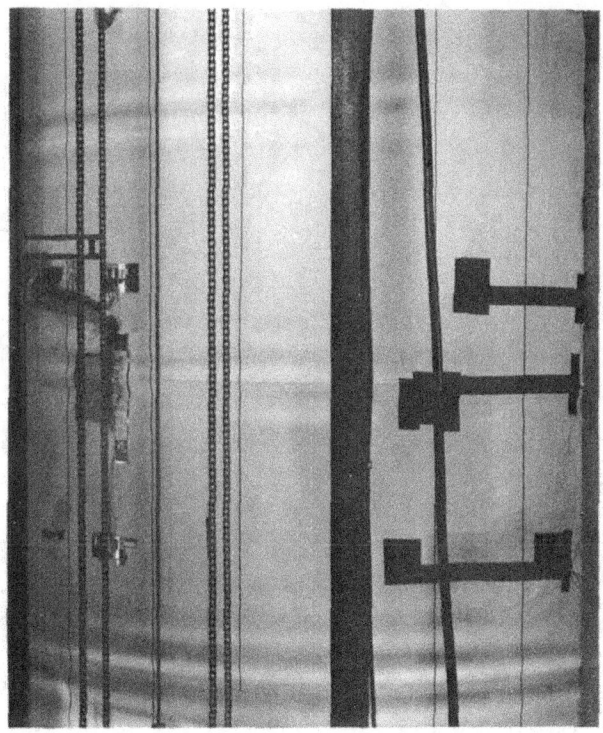

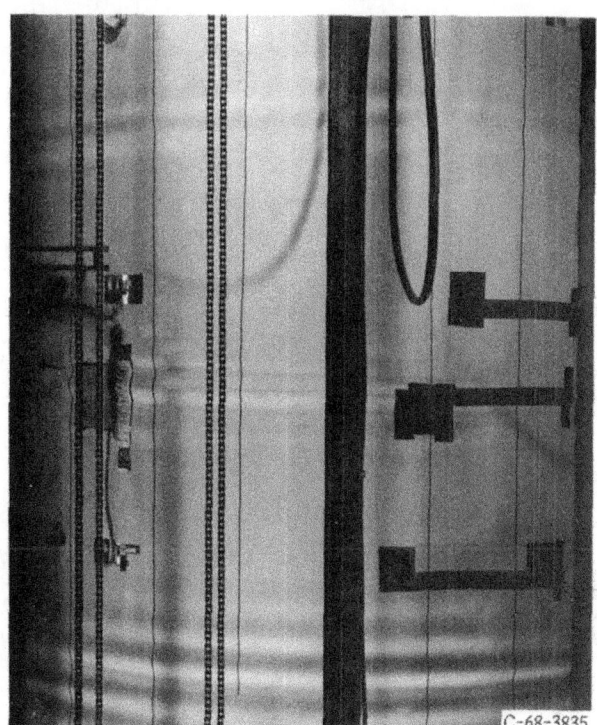

(g) Bending moment, 10.15x10⁶ inch-pounds (1146x10³ m-N).

(h) Bending moment, 10.9x10⁶ inch-pounds (1230x10³ m-N).

Figure 21. - Concluded.

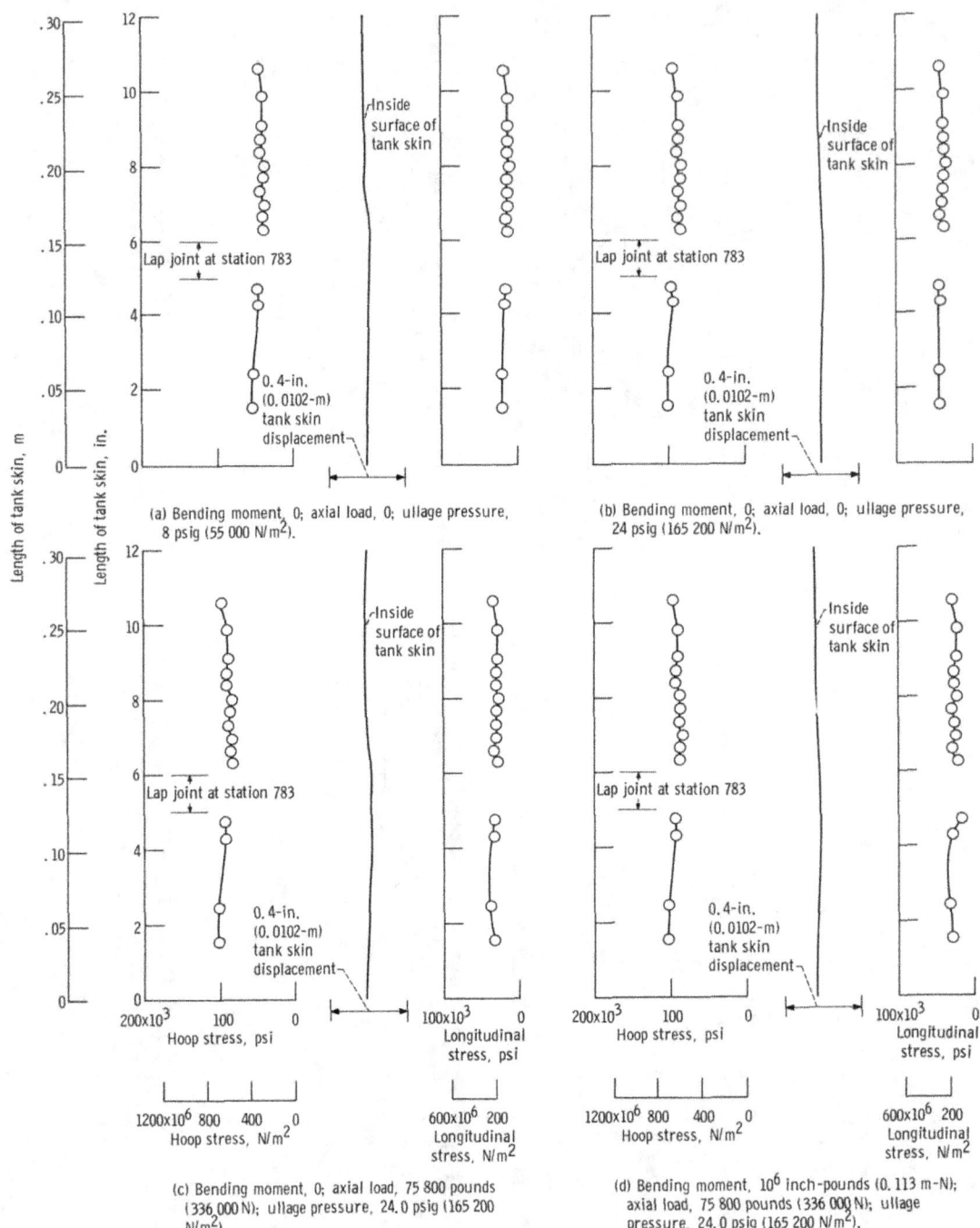

(a) Bending moment, 0; axial load, 0; ullage pressure, 8 psig (55 000 N/m²).

(b) Bending moment, 0; axial load, 0; ullage pressure, 24 psig (165 200 N/m²).

(c) Bending moment, 0; axial load, 75 800 pounds (336 000 N); ullage pressure, 24.0 psig (165 200 N/m²).

(d) Bending moment, 10⁶ inch-pounds (0.113 m-N); axial load, 75 800 pounds (336 000 N); ullage pressure, 24.0 psig (165 200 N/m²).

Figure 22. - Atlas postwrinkling-strength test. Comparison of wrinkle shape, longitudinal stress, and hoop stress at maximum compression fiber. Hydrostatic head, 213 inches (5.4 m) of water.

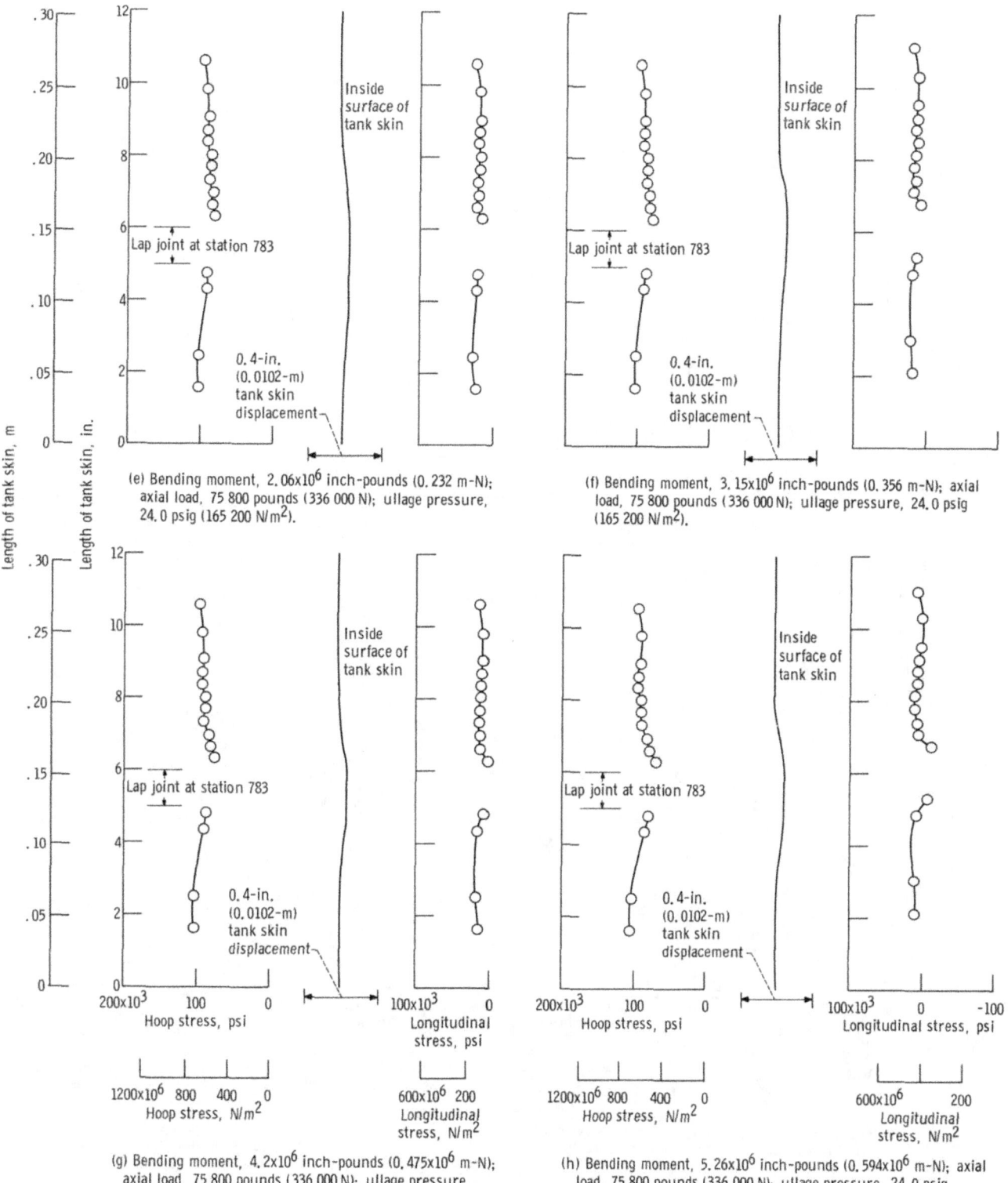

(e) Bending moment, 2.06x10⁶ inch-pounds (0.232 m-N); axial load, 75 800 pounds (336 000 N); ullage pressure, 24.0 psig (165 200 N/m²).

(f) Bending moment, 3.15x10⁶ inch-pounds (0.356 m-N); axial load, 75 800 pounds (336 000 N); ullage pressure, 24.0 psig (165 200 N/m²).

(g) Bending moment, 4.2x10⁶ inch-pounds (0.475x10⁶ m-N); axial load, 75 800 pounds (336 000 N); ullage pressure, 24.0 psig (165 200 N/m²).

(h) Bending moment, 5.26x10⁶ inch-pounds (0.594x10⁶ m-N); axial load, 75 800 pounds (336 000 N); ullage pressure, 24.0 psig (165 200 N/m²).

Figure 22. - Continued.

64

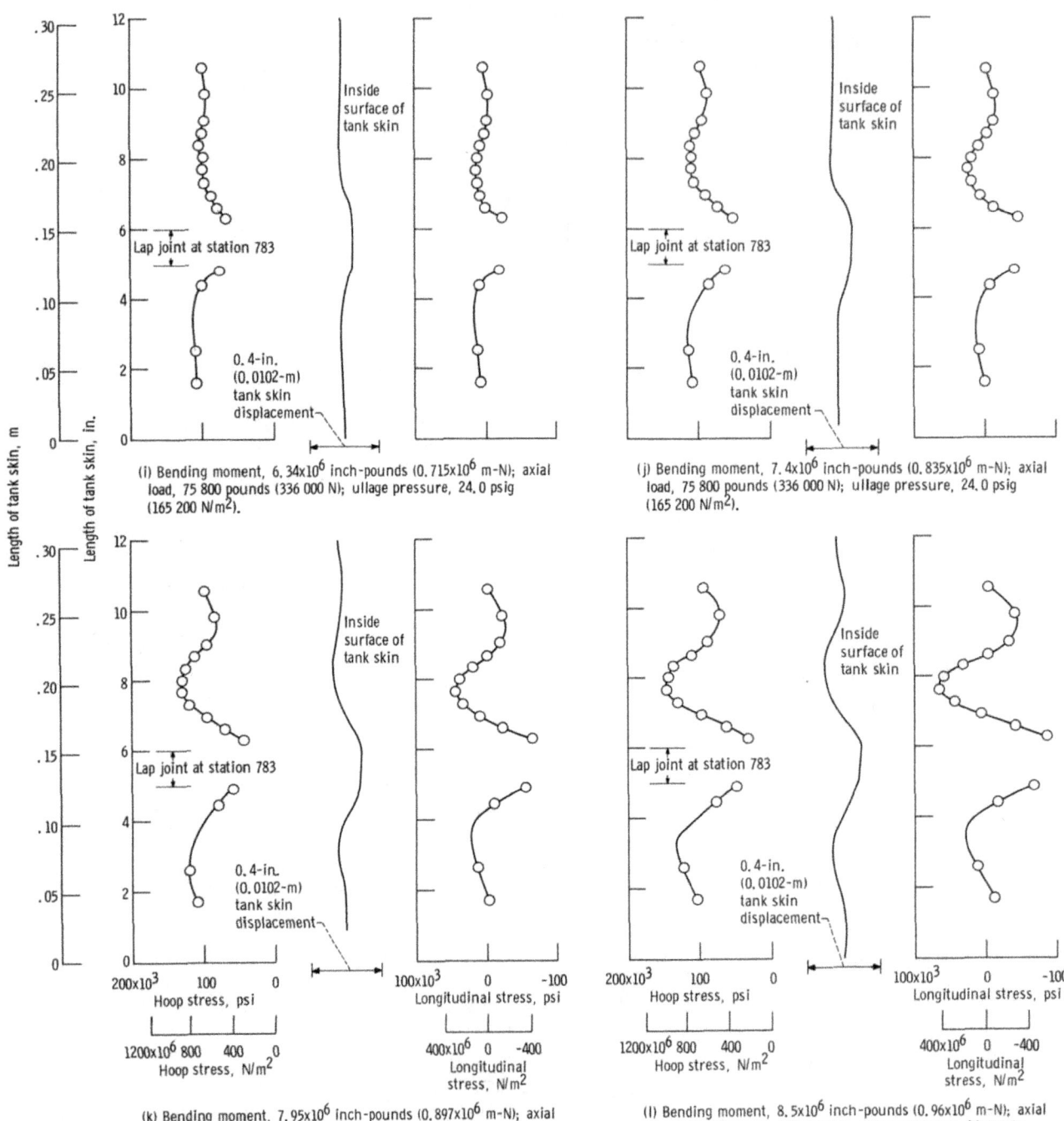

(i) Bending moment, 6.34x10⁶ inch-pounds (0.715x10⁶ m-N); axial load, 75 800 pounds (336 000 N); ullage pressure, 24.0 psig (165 200 N/m²).

(j) Bending moment, 7.4x10⁶ inch-pounds (0.835x10⁶ m-N); axial load, 75 800 pounds (336 000 N); ullage pressure, 24.0 psig (165 200 N/m²).

(k) Bending moment, 7.95x10⁶ inch-pounds (0.897x10⁶ m-N); axial load, 75 800 pounds (336 000 N); ullage pressure, 24.0 psig (165 200 N/m²).

(l) Bending moment, 8.5x10⁶ inch-pounds (0.96x10⁶ m-N); axial load, 75 800 pounds (336 000 N); ullage pressure, 24.0 psig (165 200 N/m²).

Figure 22. - Continued.

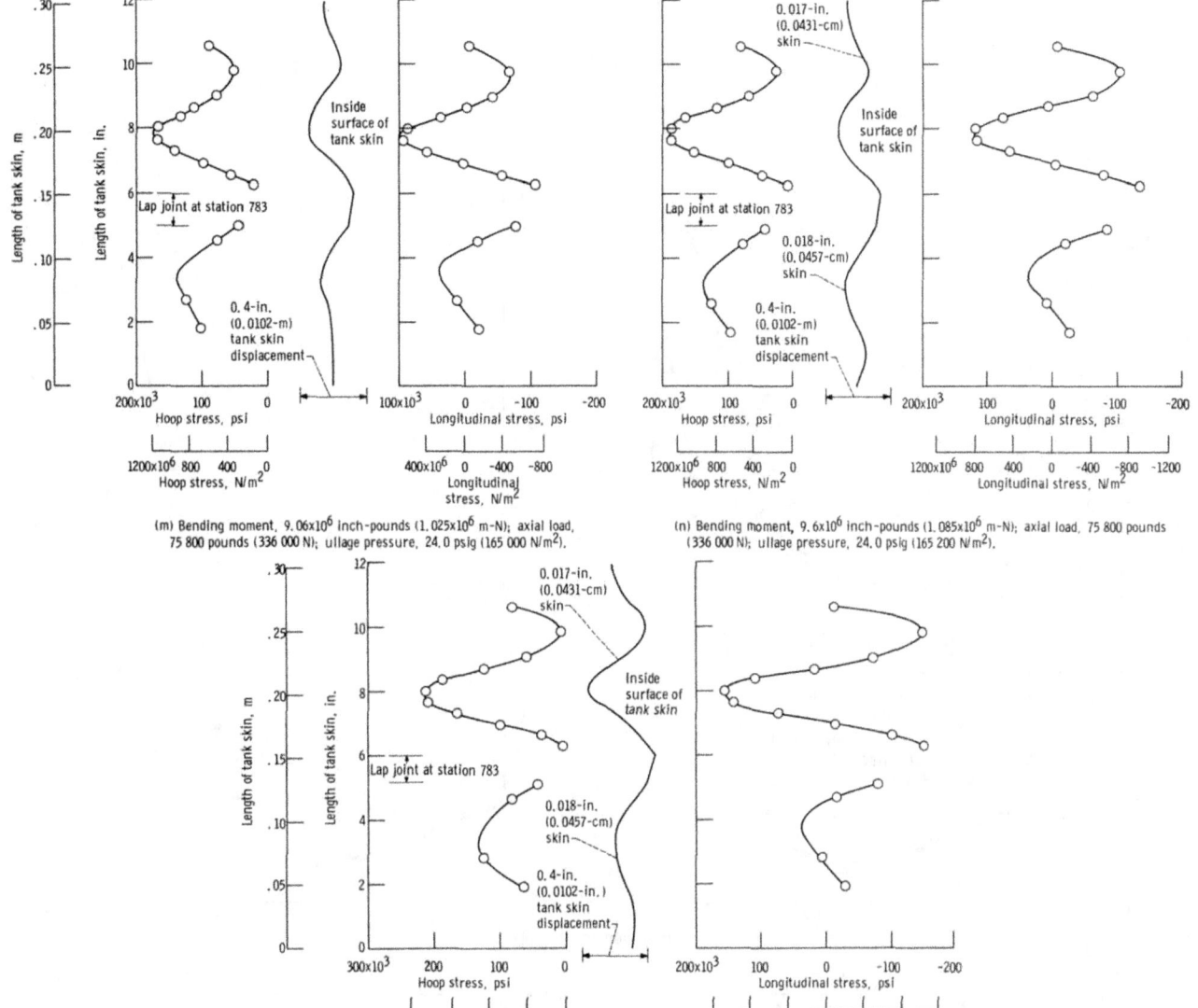

(m) Bending moment, 9.06x10⁶ inch-pounds (1.025x10⁶ m-N); axial load, 75 800 pounds (336 000 N); ullage pressure, 24.0 psig (165 000 N/m²).

(n) Bending moment, 9.6x10⁶ inch-pounds (1.085x10⁶ m-N); axial load, 75 800 pounds (336 000 N); ullage pressure, 24.0 psig (165 200 N/m²).

(o) Bending moment, 10.17x10⁶ inch-pounds (1.15x10⁶ m-N); axial load, 75 800 pounds (336 000 N); ullage pressure, 24.0 psig (165 200 N/m²).

Figure 22. - Concluded.

66

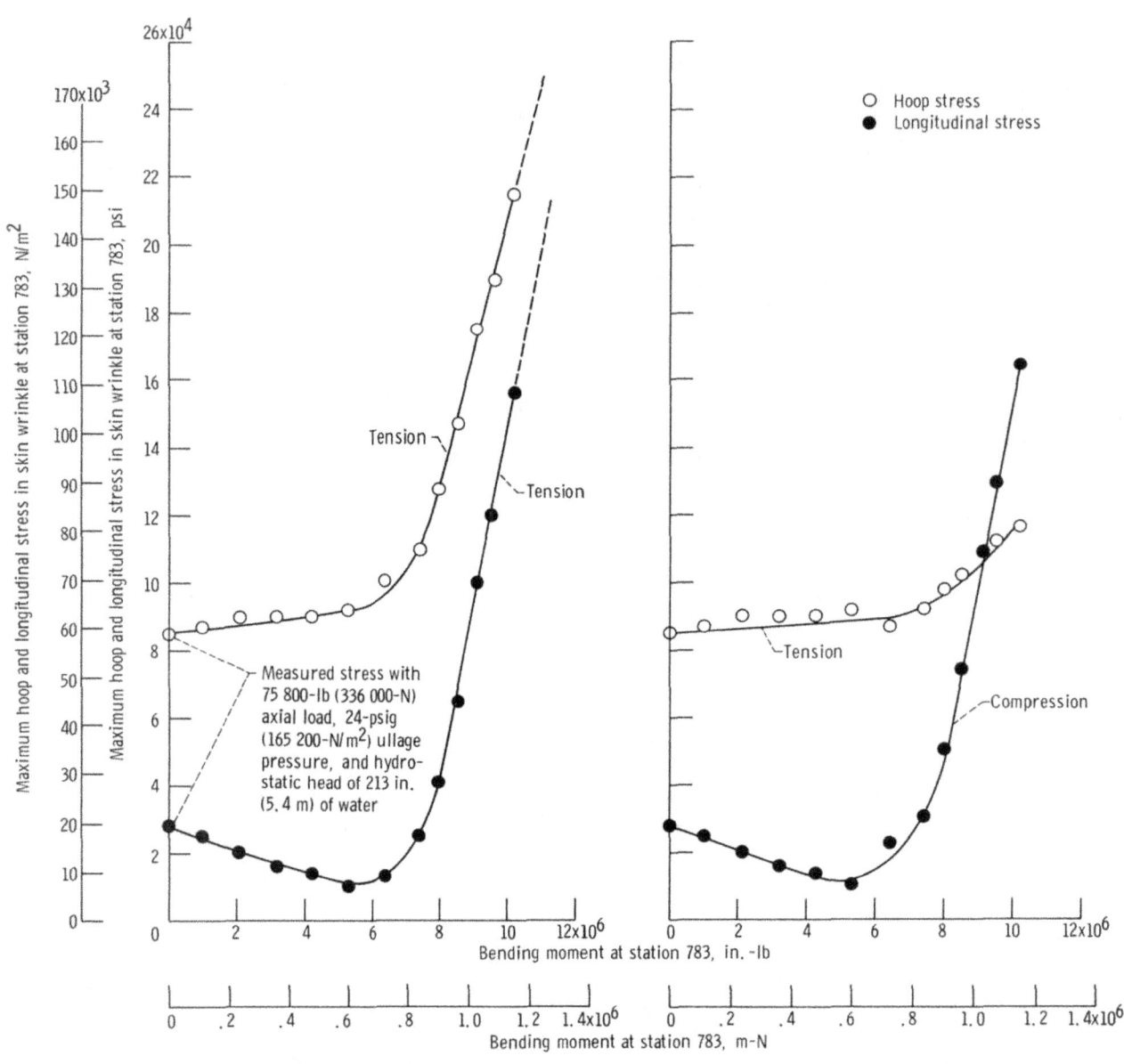

(a) Maximum hoop and longitudinal stress on outside surface of wrinkle at station 783.

(b) Maximum hoop and longitudinal stress on inside surface of buckle at station 783.

Figure 23. - Comparison of maximum stresses on inside and outside surfaces of tank skin with bending moment at station 783 (gages 1 to 8, fig. 12).

$$\sigma_{L,o} = \frac{N_c}{t} \pm \frac{6m}{t^2}$$

was used to solve for the moment in the wrinkle, where

$\sigma_{L,o}$ longitudinal stress on outside surface, psi; N/m^2

m   moment in skin in local wrinkle, (in.-lb)/in.; (m-N)/m

N_c critical wrinkling load of skin, lb/in.; N/m

t   thickness of skin, in.; m

The moment m, in turn, was used to estimate the longitudinal stress on the inside sur-face of the skin. The hoop stress on the inside surface was obtained by assuming no strain differential through the skin thickness. With this assumption, the inside-surface hoop stress can be estimated by

$$\sigma_{H,i} = \sigma_{H,o} - 2\mu\left(\sigma_{L,o} + \frac{N_c}{t}\right)$$

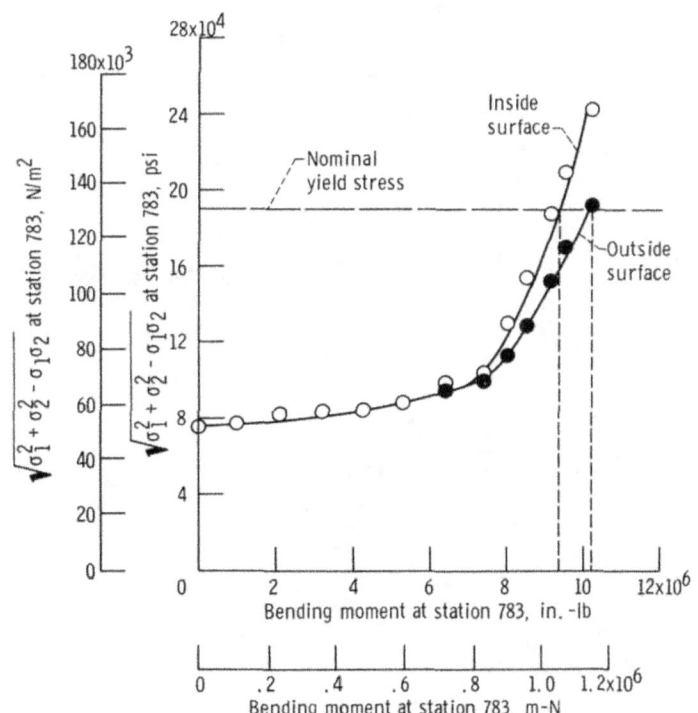

Figure 24. - Moment that produces yield in skin.
Bending moment in Atlas as function of
expression for failure from maximum distortion energy theory. Principal stresses, σ_1
and σ_2, based on data presented in figure 23.

68

where

$\sigma_{H,o}$ outside-surface hoop stress, psi; N/m^2

$\sigma_{L,i}$ inside-surface longitudinal stress, psi; N/m^2

$\sigma_{H,i}$ inside-surface hoop stress, psi; N/m^2

μ Poisson's ratio, in./in.; m/m

A derivation of the above equation is presented in appendix C.

From the stress in figures 23(a) and (b) the expression for yielding by the maximum distortion energy theory was used to obtain figure 24. This figure presents the expression for yield against applied moment. Figures 23(b) and 24 indicate yielding of the

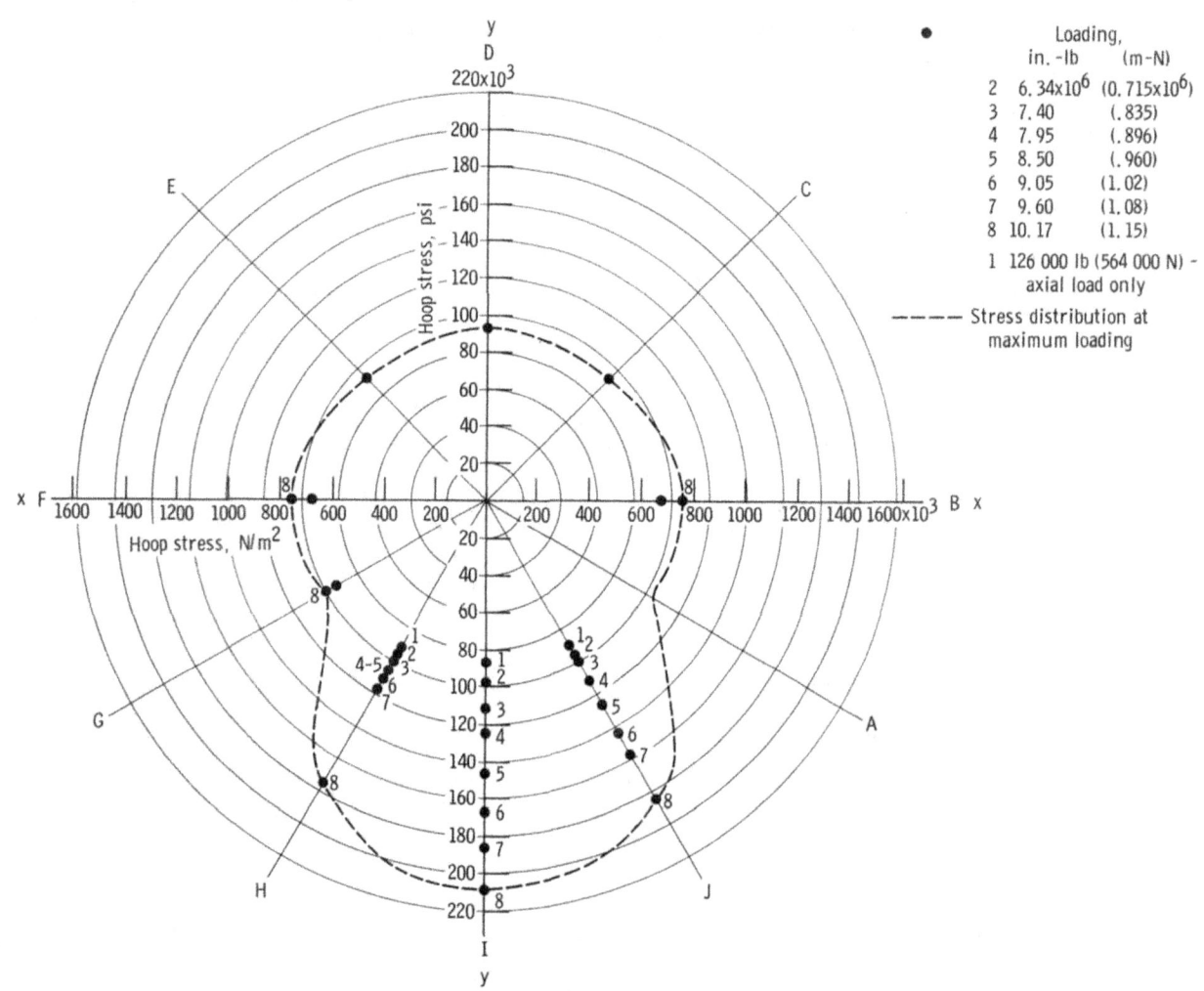

Figure 25. - Hoop stress distribution around Atlas circumference in plane of maximum wrinkle, station 780 (see fig. 11).

inner surface began at about 9.4×10^6 inch-pounds (1.06×10^6 m-N) and on the outside surface at 10.2×10^6 inch-pounds (1.15×10^6 m-N). In test 1, however, no visible sign of permanent set was evident after the application of 10.2×10^6 inch-pounds (1.15×10^6 m-N) of moment.

Hoop stress distribution around the tank circumference in the plane of the deepest wrinkle in the station 783 area is presented in figure 25. The stress is plotted at each strain-gage location, identified by letters A to J which correlate with the gage orientation on the tank shown in figure 11. The stress at gage location H is questionable in that the gages were located near a vertical joint in the tank skin. The stiff vertical joint prevented the ideal wrinkling pattern of the skin from developing until the 10.2×10^6-inch-pound (1.15×10^6-m-N) loading, at which time the vertical joint wrinkled and the stress assumed the more ideal symmetrical distribution.

At the conclusion of test 2 the deflection measurements returned to zero within the accuracy of the instrumentation. Examination of the Atlas after testing revealed that yielding of the skin had occurred in the wrinkles originating at stations 783, 812, and 841, station 812 being approximately the point of maximum compression. The yielded portions of the tank skin showed up as slight bulges. The most severe deformations occurred near station 812. The bulges protruded approximately 0.14 inch (0.00356 m) at the peak and extended, with decreasing amplitude, 70° around the tank circumference on each side of the maximum compression point. A thorough leak check of the vehicle with pressures of 15 psig (103 200 N/m^2 gage) in the fuel tank and 8 psig (55 000 N/m^2

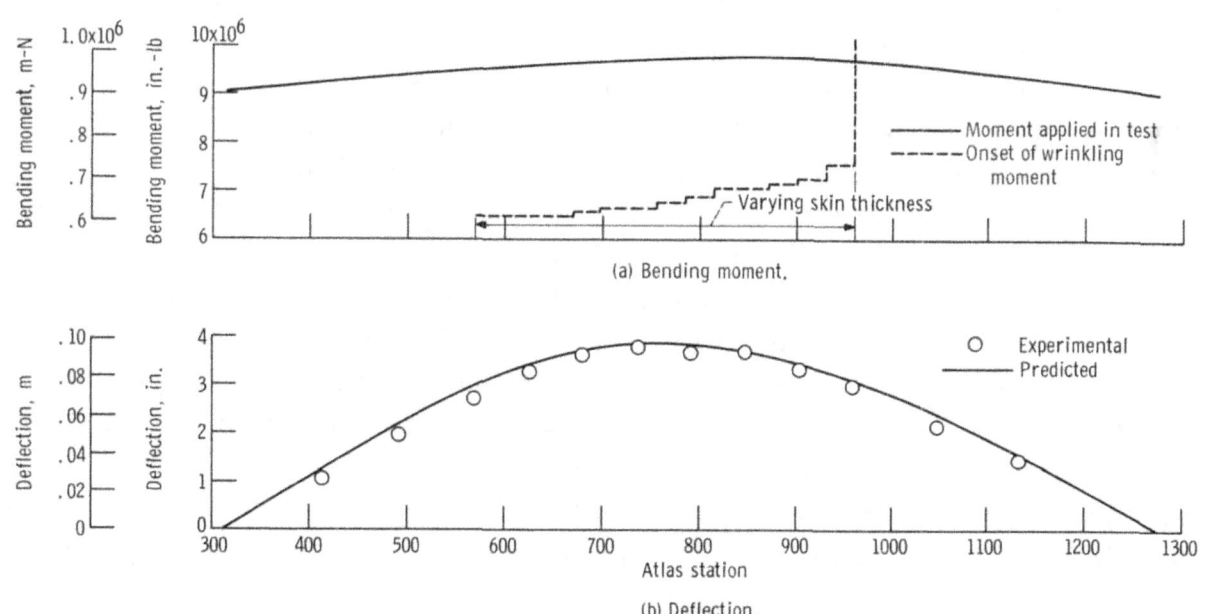

(a) Bending moment.

(b) Deflection.

Figure 26. - Bending moment and deflection as function of Atlas station - test 3. Test conditions: axial load, 115 300 pounds (511 000 N); oxidizer tank pressure, 28.5 psig (196 000 N/m^2); fuel tank pressure, 58.5 psig (403 000 N/m^2); oxidizer tank filled with water to station 540; fuel tank filled with water to station 925.

70

gage) in the oxidizer tank, using soap solution, revealed no leaks in the tank structure.

For test 3, the vehicle was rotated 40°, moving the previously yielded skin away from maximum compression and placing the long (or B-1) pod in the position of the extreme compression fiber. Test 3 was devised to investigate the bending capability of the entire vehicle and not just the cylindrical monocoque portion of the tank as in tests 1 and 2. Test 3 placed a moment over the entire Atlas and interstage adapter in excess of 9×10^6 inch-pounds (1.02×10^6 m-N), as shown in figure 26(a). The maximum load shown in figure 26 was limited by the capability of the test fixtures. This limited capability allowed an axial load of 115 300 pounds (512 000 N) instead of the 126 000 pounds (560 000 N) used for all other test 3 loadings. The measured deflection is compared to predicted deflection in figure 26(b) (see appendix A).

For typical flights at the time of maximum αQ, the predicted bending moment distribution over the Atlas-Centaur vehicle indicates that the moments developed at stations 570 and 960 are approximately 80 percent of the peak moment at station 812. The moments applied to stations 570 and 960 in test 3 were 85 and 86 percent, respectively, of the maximum moment (11.2×10^6 in.-lb, or 1.265×10^6 m-N) established in test 2 for station 812. Therefore, these test loadings show that the overall bending-strength envelope of the Atlas is compatible with the capability of the center portion (approx. station 783) of the oxidizer tank for typical Atlas-Centaur launches.

Figure 27 shows the load distribution around the tank circumference in the middle of the 0.014-inch- (0.000356-m-) thick skin just below station 570 during test 3. The curves show predicted longitudinal load against tank circumference compared with strain-gage test data. The element of the tank circumference at the 270° point received maximum compression. As the moment increased from 6.3×10^6 to 6.9×10^6 inch-pounds (0.71×10^6 to 0.78×10^6 m-N), wrinkling of the skin began, as indicated by the flat portion of each load curve in figure 27. The length of the flat portion describes the extent of wrinkling around the circumference. The curves indicate that the wrinkle angle progresses from about 52° at 6.9×10^6 inch-pounds (0.78×10^6 m-N) to 152° at 9.6×10^6 inch-pounds (1.085×10^6 m-N). These strain-gage data, in general, agree with the predictions and confirm the similar data evaluated in test 1 (fig. 16). The loading of test 3 placed a maximum tensile load of 1534 pounds per inch (268 000 N/m) and a maximum compressive load of 910 pounds per inch (159 000 N/m) on the ring at station 570. At the conclusion of test 3, close visual inspection of the test vehicle revealed no evidence of yielding or damage to the tank skins, the ring at station 570, the area around station 960 (intermediate bulkhead junction), or the interstage adapter. A soapbubble check at

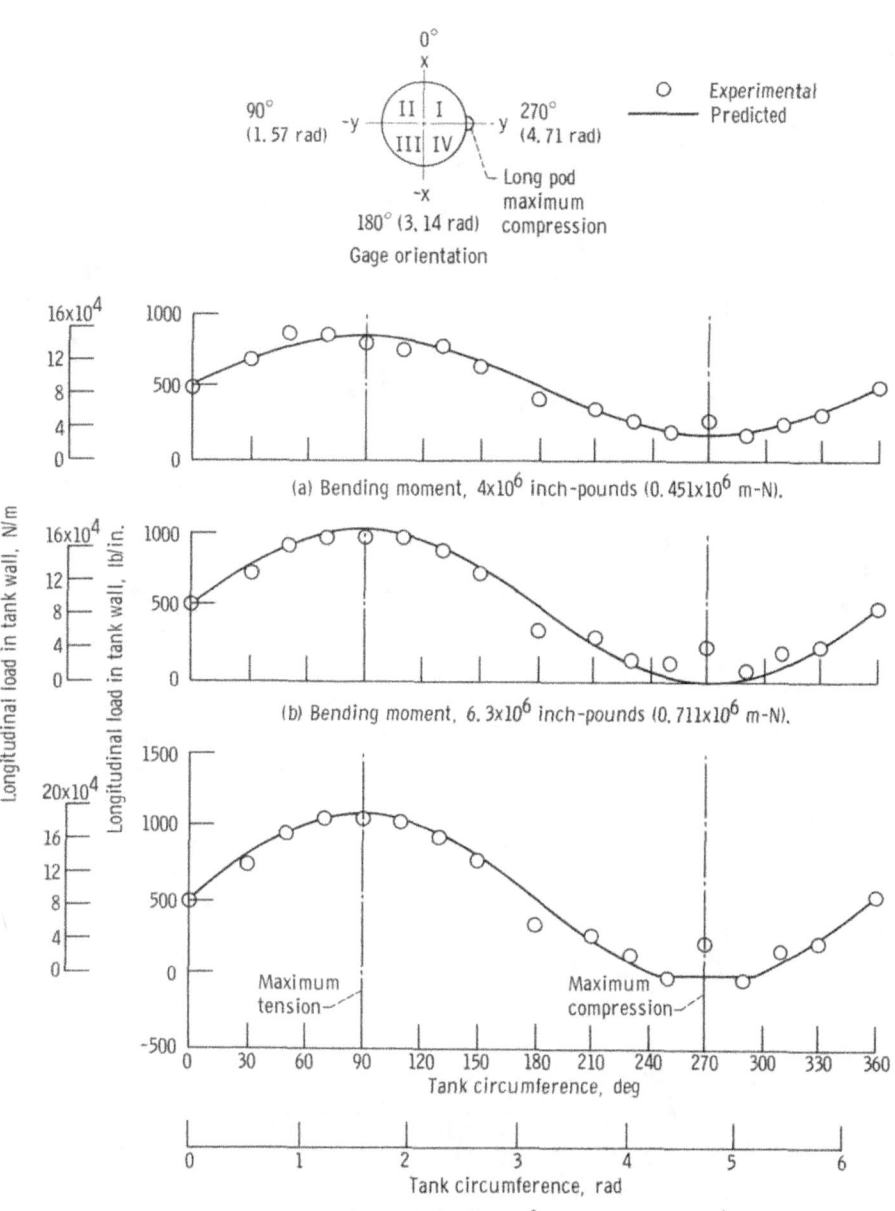

(a) Bending moment, 4x10⁶ inch-pounds (0.451x10⁶ m-N).

(b) Bending moment, 6.3x10⁶ inch-pounds (0.711x10⁶ m-N).

(c) Bending moment, 6.9x10⁶ inch-pounds (7.8x10⁶ m-N).

Figure 27. - Measured load distribution around circumference of Atlas at station 584 compared with predicted - test 3. (Flat portion of curves indicates portion of tank skin that has wrinkled.)

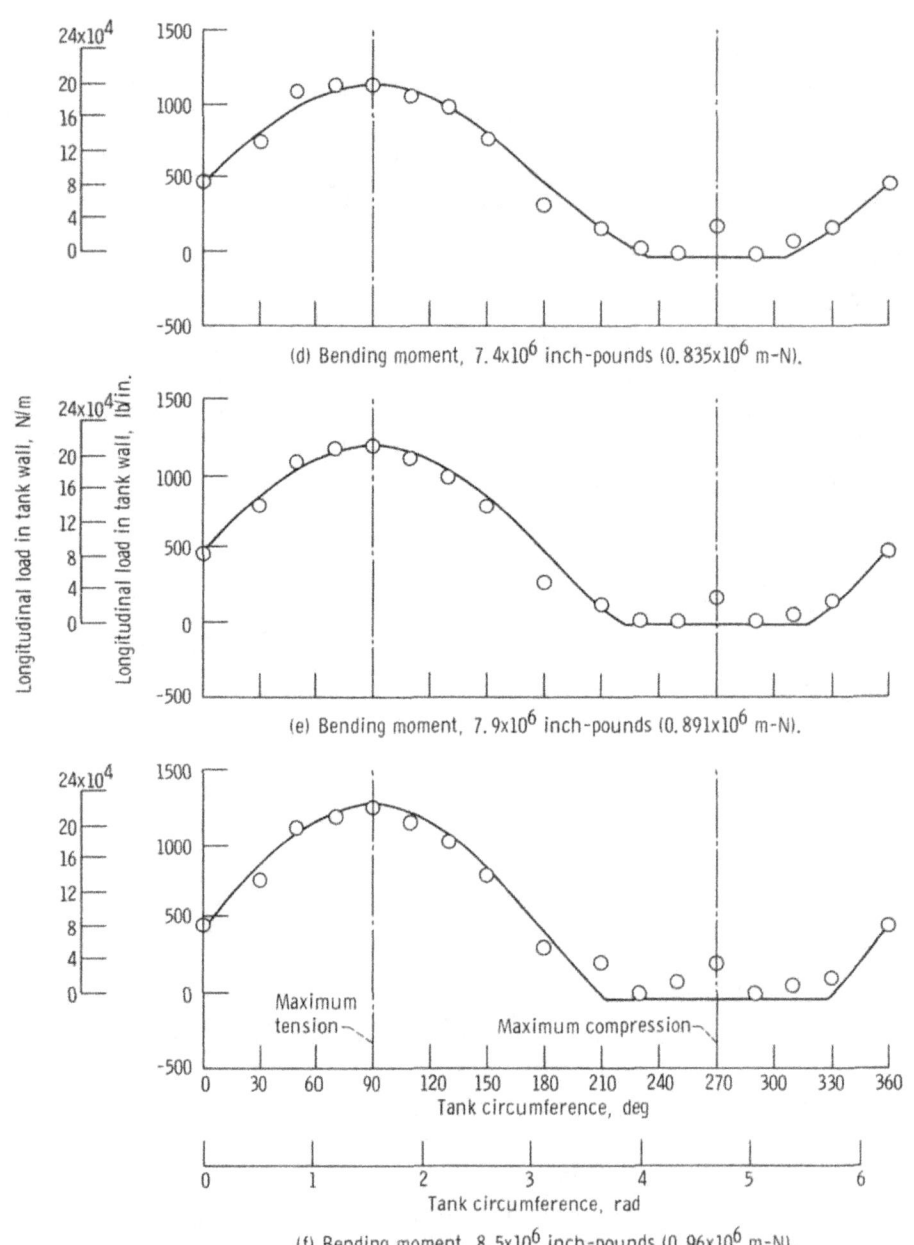

(d) Bending moment, 7.4×10^6 inch-pounds (0.835×10^6 m-N).

(e) Bending moment, 7.9×10^6 inch-pounds (0.891×10^6 m-N).

(f) Bending moment, 8.5×10^6 inch-pounds (0.96×10^6 m-N).

Figure 27. - Continued.

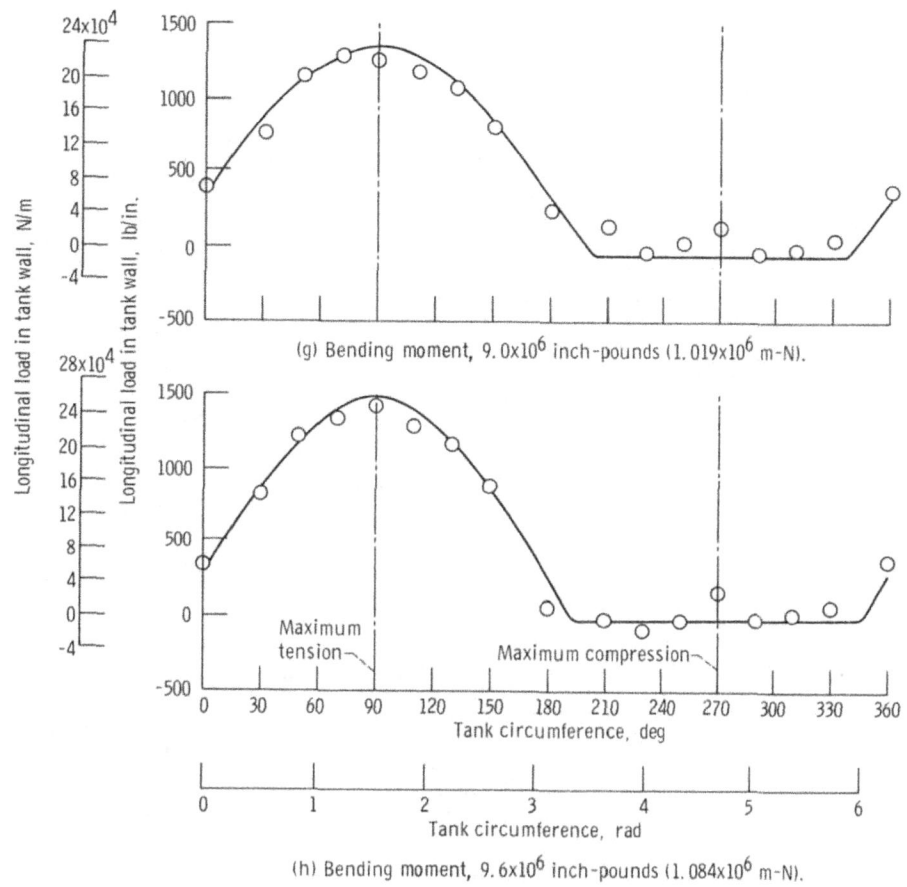

(g) Bending moment, 9.0×10^6 inch-pounds (1.019×10^6 m-N).

(h) Bending moment, 9.6×10^6 inch-pounds (1.084×10^6 m-N).

Figure 27. - Concluded.

standby pressures revealed no indication of tank leakage. Figure 28 depicts the Atlas tank under the influence of the maximum test 3 load that is shown in figure 26. The pod in the right foreground of figure 28 is receiving maximum compression. Wrinkles can be observed at each lap joint in the oxidizer tank. The severe wrinkles at station 812 are in the skin previously yielded in test 2.

Through the course of the series of three tests, the middle portion of the Atlas oxidizer tank went in and out of wrinkling eight times. Wrinkling loads were applied and removed four times in test 3, involving sections of tank skin that had previously been stressed beyond the yield point. The previously yielded condition of several tank skins, which resulted from test 2, appeared to have little effect on either the ability of the tank to sustain wrinkling loads or the predictability of the tank behavior. Figure 29 summarizes the maximum moment loadings applied to the vehicle in the test series.

Skin wrinkling that occurred near, or under, attachments to the tank wall, such as the equipment pods, appeared to have no adverse effect on the structural integrity of either the tank skin or the attachment brackets. An example of wrinkling under brackets

74

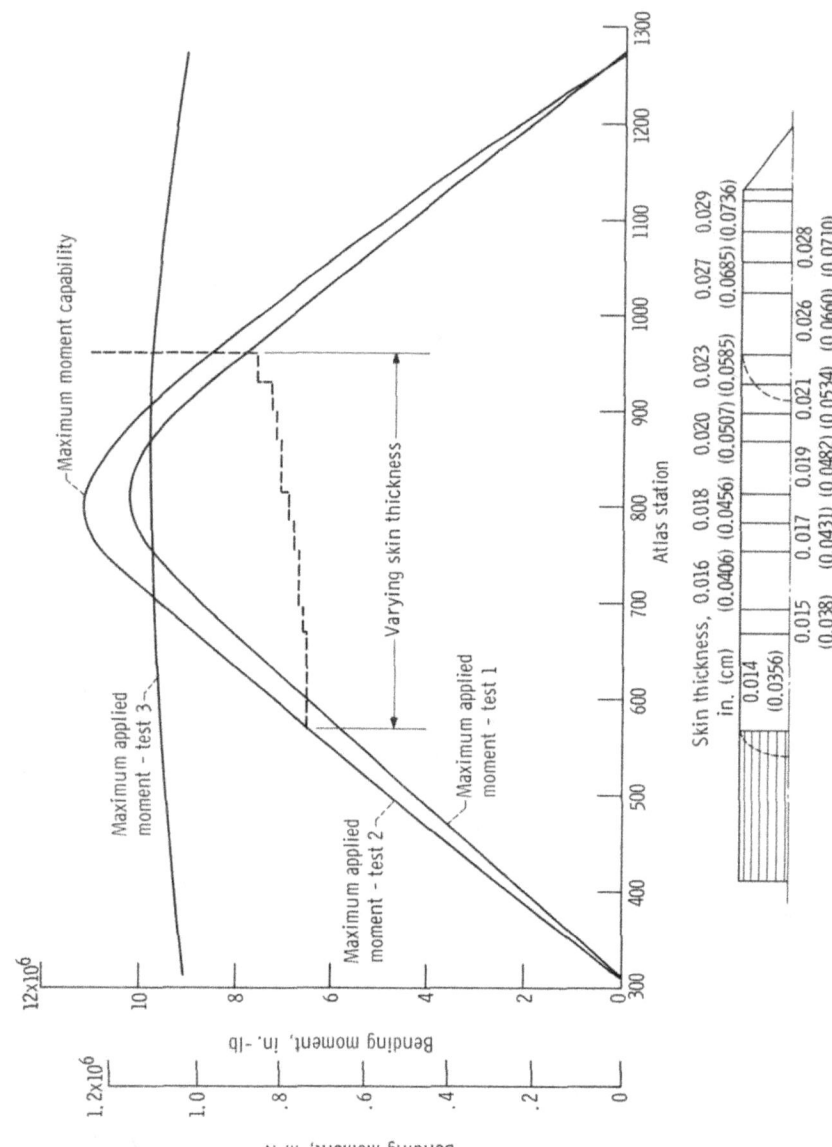

Figure 29. – Maximum applied moment as function of Atlas station.

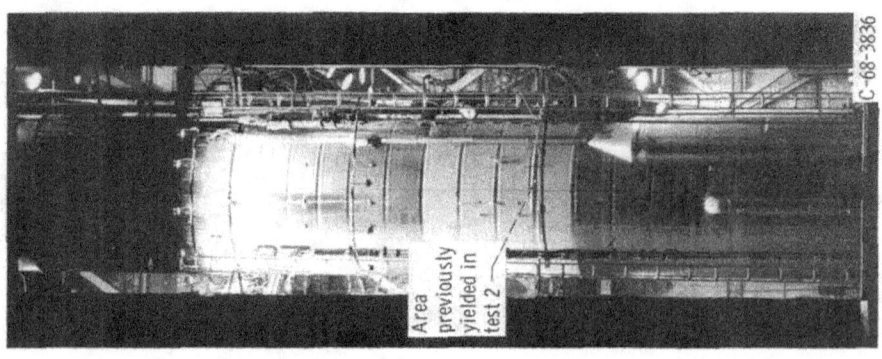

Figure 28. – Typical wrinkle pattern – test 3.

75

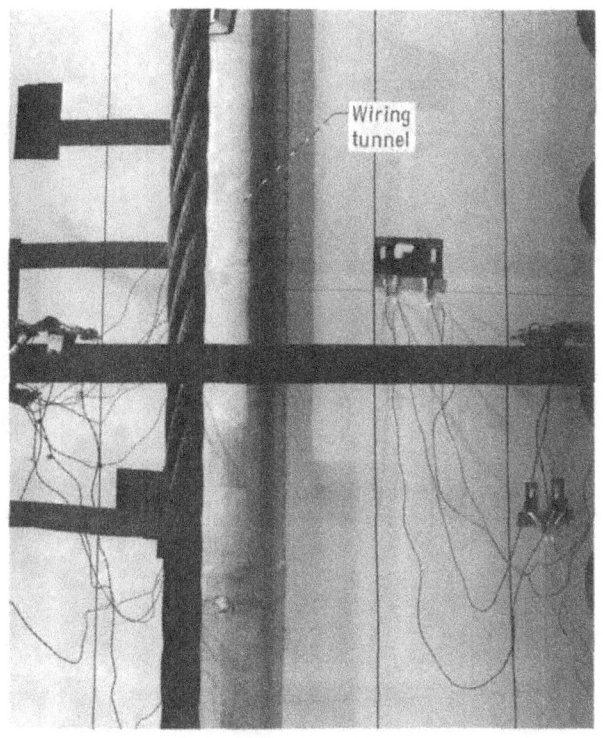

(a) Bending moment, 6.35x10^6 inch-pounds (0.716x10^6 m-N).

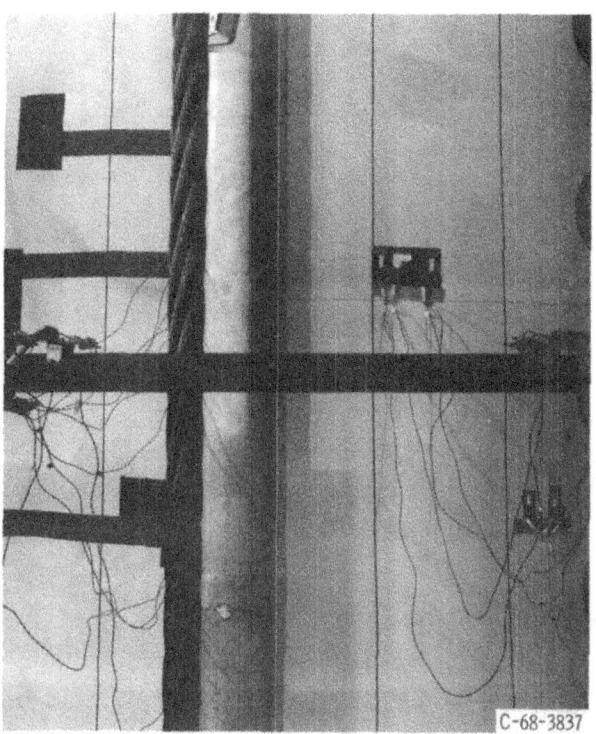

(b) Bending moment, 7.4x10^6 inch-pounds (0.835x10^6 m-N).

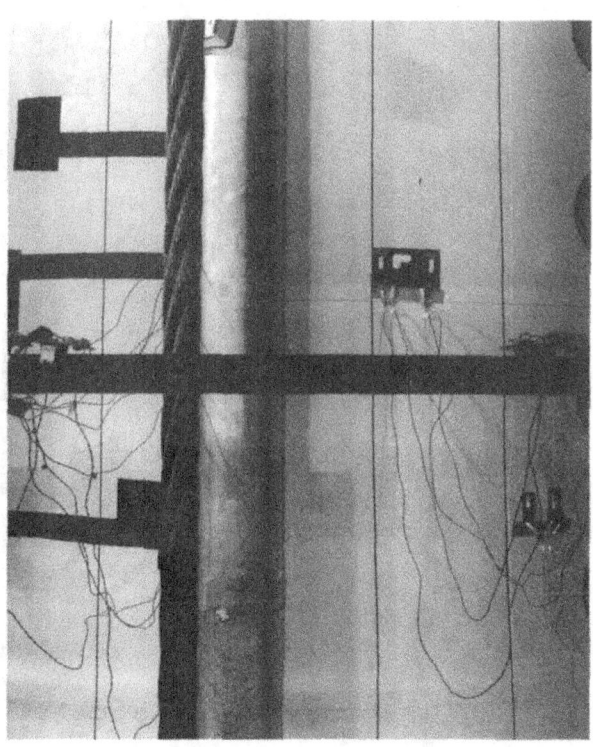

(c) Bending moment, 7.95x10^6 inch-pounds (0.898x10^6 m-N).

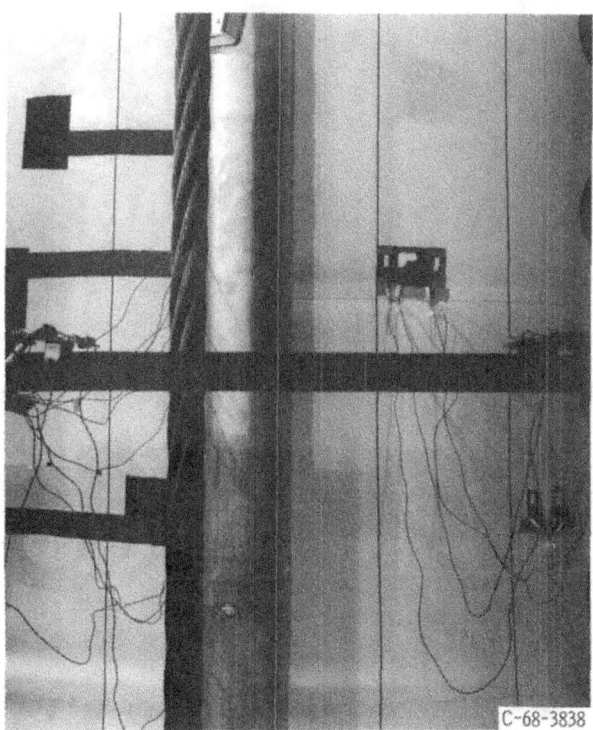

(d) Bending moment, 8.50x10^6 inch-pounds (0.96x10^6 m-N).

Figure 30. – Wrinkling under wiring tunnel – test 3.

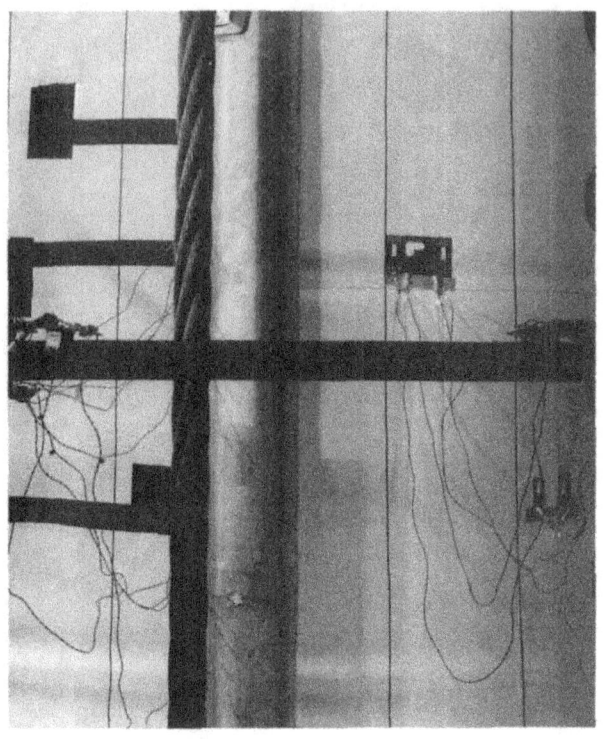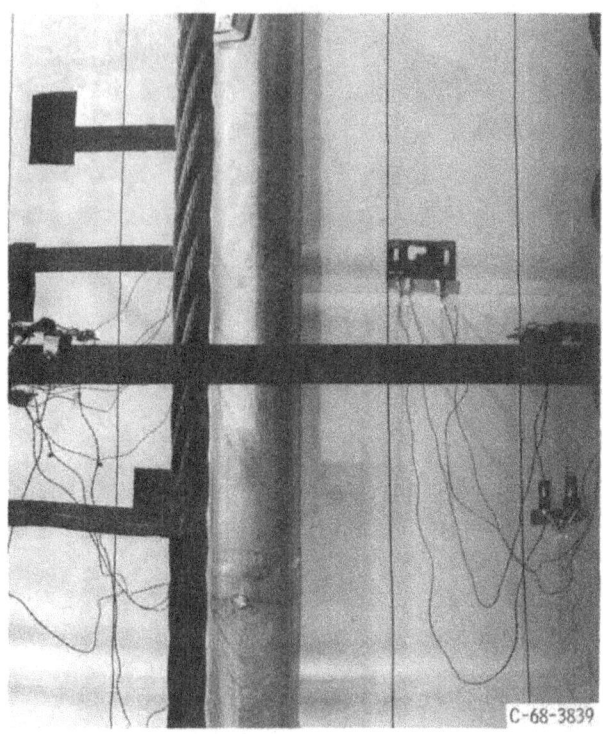

(e) Bending moment, 9.05×10^6 inch-pounds (1.023×10^6 m-N).   (f) Bending moment, 9.6×10^6 inch-pounds (1.85×10^6 m-N).

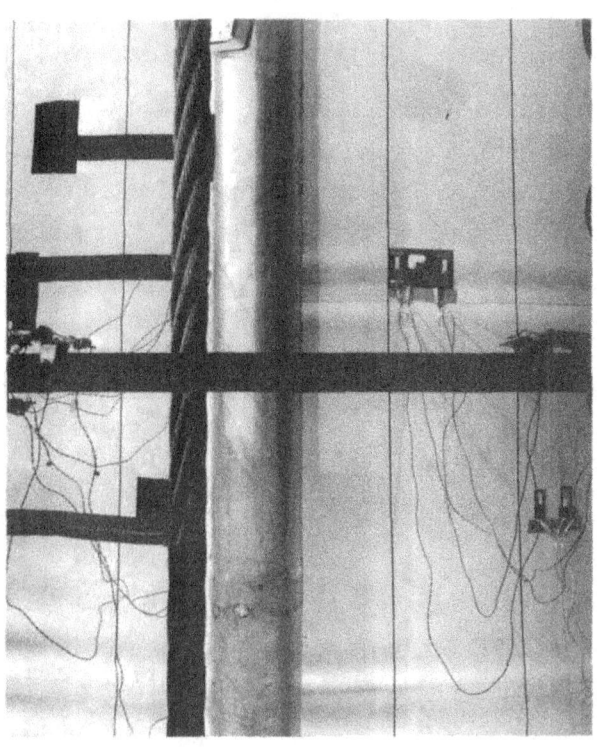

(g) Bending moment, 10.15×10^6 inch-pounds (1.15×10^6 m-N).   (h) Bending moment, 10.9×10^6 inch-pounds (1.23×10^6 m-N).

Figure 30. - Concluded.

is shown in figure 30. The reinforcing or stiffening effect of the longitudinal seams in the skins did disturb the predicted wrinkle pattern in the heavily loaded area at station 812. The vertical stiffener created by the longitudinal joint precipitated a discontinuity in the wrinkles as they progressed around the circumference of the tank. This deviation from the predicted behavior is evident in the foreground of figure 28 where the deep wrinkle at station 812 terminated above the circumferential lap joint at the longitudinal joint and shifted down below the lap joint to the less stiff portion of the tank.

CONCLUDING REMARKS

The results of this series of tests of the Atlas booster can be summarized as follows:

1. Postwrinkling behavior as predicted by reference 1 and experienced on small-scale model tests, in general, is valid for full-scale Atlas tanks.

2. Under the conditions of this investigation, the moment capability of the pure monocoque portion of the Atlas oxidizer tank is at least 11.2×10^6 inch-pounds (1.265×10^6 m-N).

3. Under the conditions of these tests, the Atlas vehicle can sustain moments of at least 9×10^6 inch-pounds (1.02×10^6 m-N) at any station on the structure and maintain structural integrity.

4. The Atlas vehicle will sustain repeated loading beyond the onset of skin wrinkling.

5. The wiring tunnel and its method of attachment have a negligible effect on the skin-wrinkling pattern and conversely, the skin wrinkling has no detrimental effect on the tunnel.

6. Wrinkling of the tank wall initiates at the skin lap joints. However, the overall wrinkling behavior of the tank closely followed that predicted for a smooth cylinder without joints.

A method of analysis for predicting stress in the local wrinkle areas is presented in reference 11. A comparison of analytical results and the experimental findings of this test program is incorporated in that reference.

Lewis Research Center,
   National Aeronautics and Space Administration,
      Cleveland, Ohio, November 6, 1968,
         491-05-00-22.

78

APPENDIX A

METHOD OF PREDICTING BEAM DEFLECTIONS

The following procedure for obtaining the predicted deflection of the Atlas beam was established for the Plum Brook tests. The Atlas was considered as a pin-supported beam column, idealized as shown in sketch a where P is the applied axial load, R_T

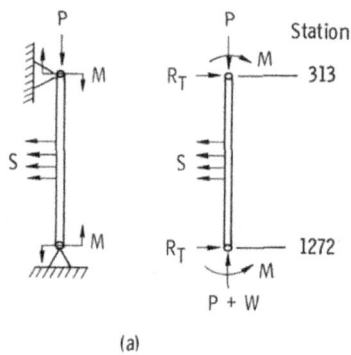

(a)

is the reaction to the tower, M is the applied moment couple, S is the applied shear load, and W is the vehicle weight.

For tests 1 and 2 the applied moments M were zero. For test 3 the applied shear loads S were zero. As the shear S or moment M is applied, the beam deflects. These deflections induce additional moments (secondary moments) through the action of the applied axial load P and the vehicle weight W. The secondary moments were obtained as indicated in sketch b where y is the deflection induced by applied shear S or moment M, w indicates points where vehicle weight is considered to be concentrated, r_T is the reaction to the tower induced by deflections, P is the axial load, and W is the vehicle weight.

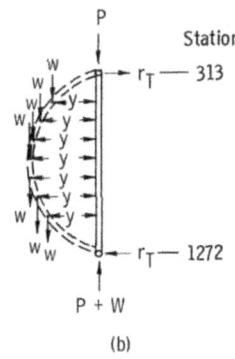

(b)

The conjugate beam method of calculating beam deflections was used to obtain the primary deflections represented as y in sketch b. With the column loaded as in sketch a, secondary moments were obtained at each w location by summing moments about each point, the moments being a function of P, R_T, and W. The secondary moments thus obtained were added to the primary moments from the applied loads, and the beam deflections were recalculated with the conjugate beam concept, by using this total moment distribution. The deflections presented in this report were obtained with iterations of this procedure.

Before wrinkling of the tank skin the deflection is linear with load. After the onset of wrinkling the deflection becomes nonlinear. In the nonlinear region the stiffness is obtained by using an effective moment of inertia I_{eff}, ($I_{eff} = tR^3 [\theta - \sin \theta \cos \theta]$) that is a function of the extent of wrinkling around the tank circumference. The origin of I_{eff} is presented in appendix B.

To monitor the magnitude of the secondary moment during testing, a ratio of secondary moment to deflection was established with the above procedure. With the assumption that the secondary moment remains proportional to deflection, this ratio was used with the deflections measured during testing to determine the total bending moment on the vehicle at any time in the loading sequence.

APPENDIX B

BEAM ANALOGY EQUATIONS FOR POSTWRINKLING STRENGTH

A relatively simple analysis was used for predicting the strength of the Atlas test vehicle. The analysis outlined herein is presented in greater detail in reference 1.

The bending stress distribution is assumed to be linear with the applied bending moment in the unwrinkled region and constant in the wrinkled region. The pressurized cylinder (sketch c) resists an external axial load P_a and an external bending moment

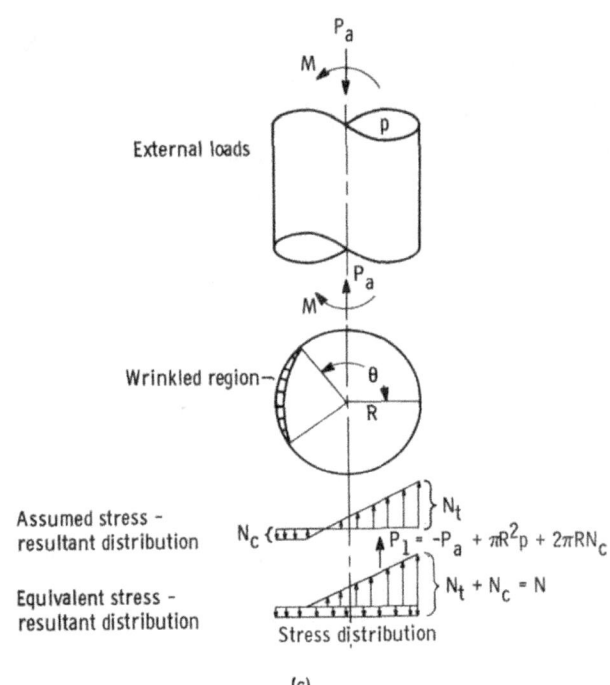

(c)

M. The bending moment is resisted by the load varying from zero at angle θ to a maximum of N per unit length. The pressure, axial load, and uniform critical wrinkling load N_c have no bending moment about the longitudinal centerline. Integrating the values for the triangular distribution yields the following equation:

$$P_1 = \frac{2NR(\sin \theta - \theta \cos \theta)}{1 - \cos \theta} \tag{1}$$

Summing the bending moment of the N distribution about the cylinder centerline gives

$$M = \frac{NR^2(\theta - \sin\theta \cos\theta)}{1 - \cos\theta} \qquad (2)$$

The critical wrinkling load N_c for the various skin thicknesses is obtained with the general equation

$$N_c = \frac{1}{\sqrt{3(1 - \mu^2)}} \left(\frac{Et^2}{R}\right) \qquad (3)$$

for a thin-walled unpressurized cylinder (ref. 3). These equations can be used to determine the stress or load distribution around the tank circumference. The total axial tensile force P_1 can also be expressed as

$$P_1 = -P_a + \pi R^2 p + 2\pi R N_c \qquad (4)$$

With this expression and the preceding equations, the extent of wrinkling around the tank circumference can be found as a function of applied moment, axial load, and internal ullage pressure.

To define the nonlinear stiffness of the wrinkled tank, an effective moment of inertia I_{eff} that is a function of the extent of skin wrinkling can be used. This effective moment of inertia is derived in the following manner. As the stress N_t varies in a distance $R(1 - \cos\theta)$, an elemental length of cylinder ΔS bends through an angle $\Delta\varphi$ (sketch d). With the assumption that plane sections remain plane, these equations can be obtained from sketch d.

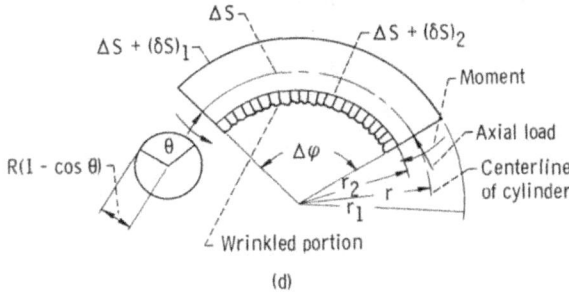

(d)

82

$$\Delta S + (\delta S)_1 = r_1 \Delta \varphi \qquad (5a)$$

$$\Delta S + (\delta S)_2 = r_2 \Delta \varphi \qquad (5b)$$

$$r_1 - r_2 = R(1 - \cos\ \theta)$$

Thus,

$$r_1 = R(1 - \cos\ \theta) + \frac{\Delta S + (\delta S)_2}{\Delta \varphi}$$

and

$$\Delta S + (\delta S)_1 = \Delta \varphi R(1 - \cos\ \theta) + \Delta S + (\delta S)_2$$

giving

$$\Delta \varphi = \frac{(\delta S)_1 - (\delta S)_2}{R(1 - \cos\ \theta)} \qquad (6)$$

Since $\epsilon_1 = N_t/Et$ and $\epsilon_2 = N_c/Et$

$$(\delta S)_1 - (\delta S)_2 = \epsilon_1 \Delta S - \epsilon_2 \Delta S$$

or

$$(\delta S)_1 - (\delta S)_2 = \frac{\Delta S}{Et}\ (N_t + N_c) \qquad (7)$$

Now, substituting equation (7) into equation (6) yields

$$\Delta \varphi = \Delta S\ \frac{N_t + N_c}{EtR(1 - \cos\ \theta)}$$

or

$$\frac{\Delta \varphi}{\Delta S} = \frac{N_t + N_c}{EtR(1 - \cos\ \theta)} \qquad (8)$$

83

$$\frac{\Delta \varphi}{\Delta S} = \frac{M}{EI_{eff}} \tag{9}$$

(ref. 12) where I_{eff} is the effective moment of inertia. Substituting equation (8) into equation (9) and adjusting terms give

$$I_{eff} = \frac{MtR(1 - \cos \theta)}{N_t + N_c} \tag{10}$$

Now substituting equation (2) into equation (10) and noting $N_t + N_c = N$ gives

$$I_{eff} = tR^3(\theta - \sin \theta \cos \theta) \tag{11}$$

This expression for moment of inertia can be used for deflection computations of pressure stabilized vehicles in the nonlinear (wrinkled) range.

APPENDIX C

DERIVATION OF EQUATION FOR INSIDE-SURFACE HOOP STRESS

The origin of the equation $\sigma_{H,i} = \sigma_{H,o} - 2\mu(\sigma_{L,o} + N_c/t)$ for hoop stress inside the tank is illustrated with the following sketch (tension considered positive):

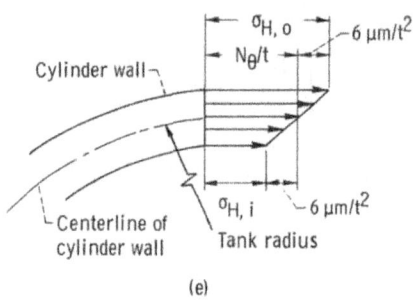

(e)

where

$\sigma_{H,i}$ inside-surface hoop stress, psi; N/m^2

$\sigma_{H,o}$ outside-surface hoop stress, psi; N/m^2

$\sigma_{L,i}$ inside-surface longitudinal or bending stress, psi; N/m^2

$\sigma_{L,o}$ outside-surface longitudinal or bending stress, psi; N/m^2

N_θ tensile hoop load in skin from internal pressure, lb/in.; N/m

m bending moment in wrinkle, (in.-lb)/in.; (m-N)/m

μ Poisson's ratio, in./in.; m/m

t skin thickness, in.; m

N_c critical wrinkling load for skin, in.-lb; (m-N)

The principal stresses can be expressed as

$$\sigma_{L,i} = -\frac{6m}{t^2} - \frac{N_c}{t}$$

$$\sigma_{L,o} = \frac{6m}{t^2} - \frac{N_c}{t}$$

$$\sigma_{H,i} = \frac{N_\theta}{t} - \frac{\mu 6m}{t^2}$$

$$\sigma_{H,o} = \frac{N_\theta}{t} + \frac{\mu 6m}{t^2}$$

Values of $\sigma_{H,o}$ and $\sigma_{L,o}$ were obtained with strain gages and used to obtain N_θ/t as follows:

$$\frac{N_\theta}{t} = \sigma_{H,o} - \frac{\mu 6m}{t^2}$$

but

$$\sigma_{L,o} = \frac{6m}{t^2} - \frac{N_c}{t}$$

Therefore,

$$\frac{N_\theta}{t} = \sigma_{H,o} - \mu\left(\sigma_{L,o} + \frac{N_c}{t}\right)$$

Thus,

$$\sigma_{H,i} = \frac{N_\theta}{t} - \mu\left(\sigma_{L,o} + \frac{N_c}{t}\right)$$

$$= \sigma_{H,o} - \mu\left(\sigma_{L,o} + \frac{N_c}{t}\right) - \mu\left(\sigma_{L,o} + \frac{N_c}{t}\right)$$

$$= \sigma_{H,o} - 2\mu\left(\sigma_{L,o} + \frac{N_c}{t}\right)$$

REFERENCES

1. Peery, David J.: Post-Buckling Strength of a Pressurized Cylinder. Rep. GD/A-DDG-64-024A, General Dynamics/Astronautics (NASA CR-54802), Oct. 16, 1964.

2. Jahsman, W. E.: Combined Bending and Compression of a Pressurized Circular Cylindrical Membrane Column. J. Eng. Industry, vol. 87, no. 3, Aug. 1965, pp. 372-378.

3. Weingarten, V.: Effects of Internal Pressure on the Buckling of Circular Cylindrical Shells Under Bending. Rep. STL 7102-0033-RU-000, EM 11-12, Space Technology Labs., TRW, Inc., 1961.

4. Anon.: Buckling of Thin-Walled Circular Cylinders. NASA SP-8007, 1965.

5. McComb, Harvey G., Jr.; Zender, George W.; and Mikulas, Martin M., Jr.: The Membrane Approach to Bending Instability of Pressurized Cylindrical Shells. Collected Papers on Instability of Shell Structures. NASA TN D-1510, 1962, pp. 229-237.

6. Seide, P.; Weingarten, V. I.; and Morgan, E. J.: Development of Design Criteria for Elastic Stability of Thin Shell Structures. Rep. STL/TR-60-0000-19424, Space Technology Labs., TRW, Inc., 1960.

7. Suer, Herbert S.; Harris, Leonard A.; Skene, William T.; and Benjamin, Roland J.: The Bending Stability of Thin-Walled Unstiffened Circular Cylinders Including Effects of Internal Pressure. J. Aeron. Sci., vol. 25, no. 5, May 1958, pp. 281-287.

8. Leaumont, Walter J., Jr.: Collapse Tests of Pressurized Membrane-like Circular Cylinders for Combined Compression and Bending. NASA TN D-2814, 1965.

9. Zender, George W.: The Bending Strength of Pressurized Cylinders. J. Aerospace Sci., vol. 29, no. 3, Mar. 1962, pp. 362-363.

10. Miller, Robert P.; and Gerus, Theodore: Bending Strength of a Large Thin-Walled Pressure-Stabilized Cylinder Beyond Onset of Compressive Skin Wrinkling. NASA TM X-1311, 1966.

11. Greenbaum, G. A.; and Conroy, D. C.: Post-Wrinkling Analysis of Highly Pressurized Cylindrical and Conical Shells of Revolution Subjected to Bending Loads. Vol. I and II. TRW, Inc., Oct. 15, 1967.

12. Shanley, F. R.: Strength of Materials. McGraw-Hill Book Co., Inc., 1957.

EXPERIMENTAL LATERAL BENDING DYNAMICS OF THE

ATLAS-CENTAUR-SURVEYOR LAUNCH VEHICLE

By Robert P. Miller and Theodore F. Gerus

Lewis Research Center
Cleveland, Ohio

NATIONAL AERONAUTICS AND SPACE ADMINISTRATION

For sale by the Clearinghouse for Federal Scientific and Technical Information
Springfield, Virginia 22151 — CFSTI price $3.00

ABSTRACT

A series of full-scale dynamic tests was conducted to determine the lateral bending dynamics of the Atlas-Centaur-Surveyor launch vehicle. The tests were made at several simulated flight times to obtain a good experimental definition of the vehicle dynamic response in the bending mode frequency range. Parameters measured were bending mode frequencies, damping, mode shapes, and linearity. The bending mode frequencies measured ranged from 2.59 to 31.9 Hz. Damping measured averaged about 2.0 percent of critical. Mode shapes were well defined and continuous in the lower-frequency modes but indicated discontinuities in the higher modes. The modal response was very linear over the excitation range chosen, and only small differences between pitch and yaw axes were seen   Phase relations of various response points for three modes were also measured.

EXPERIMENTAL LATERAL BENDING DYNAMICS OF THE

ATLAS-CENTAUR-SURVEYOR LAUNCH VEHICLE

by Robert P. Miller and Theodore F. Gerus

Lewis Research Center

SUMMARY

A series of full-scale dynamic tests was conducted to determine the lateral bending dynamics of the Atlas-Centaur-Surveyor launch vehicle. The tests were made at several simulated flight times to obtain a good experimental definition of the vehicle dynamic response in the bending mode frequency range. Parameters measured were bending mode frequencies, damping, mode shapes, and linearity.

The bending mode frequencies measured ranged from 2.59 to 31.9 hertz. Damping measured averaged about 2.0 percent of critical. Mode shapes were well defined and continuous in the lower-frequency modes but indicated discontinuities in the higher modes. The modal response was very linear over the excitation range chosen, and only small differences between pitch and yaw axes were seen. Phase relations of various response points for three modes were also measured.

INTRODUCTION

The design of the Atlas-Centaur launch vehicle has evolved as a relatively long, thin-walled pressurized cylinder (fig. 1). The control of this structure in flight results in the interaction of the vehicle lateral bending modes with the flight control system. Such coupling exists because the control system attitude sensors detect angular position and rates caused by structural deformations. In the case of the Atlas-Centaur, the lower-frequency bending modes are within the control system response range for a large portion of powered flight. The stability of the vehicle control, therefore, requires examination of the corresponding closed-loop modes of the vehicle structure and the control system. A schematic of the control system is shown at the top of the next page.

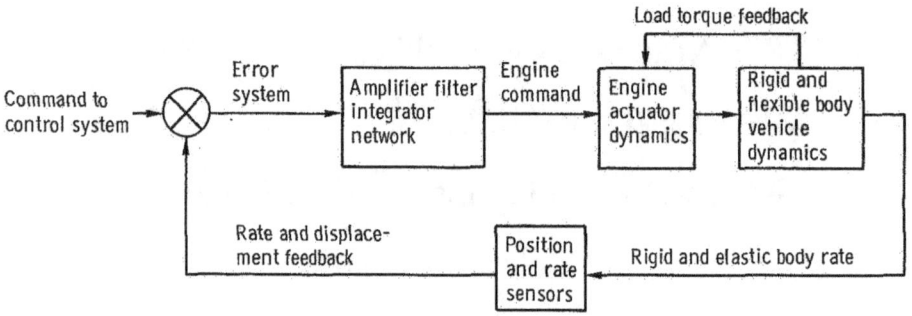

It was the purpose of the lateral dynamics tests to experimentally determine the frequencies, mode shapes, and damping characteristics of the Atlas-Centaur-Surveyor at various simulated flight times in order that accurate elastic modal data could be incorporated in the elastic stability analysis as well as in the analysis for loads and clearance. The complexity of the Atlas-Centaur structure, with its many discontinuities and branches, makes analytical predictions difficult. A test of the lateral bending dynamics of the vehicle was made to improve the analysis techniques and/or to gain confidence in these techniques.

APPARATUS AND PROCEDURE

The tests were conducted on a full-scale, production-type Atlas-Centaur vehicle and a dynamic model of the Surveyor spacecraft. The vehicle was suspended on soft springs to simulate the free-free condition of flight and excited laterally at the gimbal plane with an electrodynamic shaker. Natural frequencies, mode shapes, and damping characteristics of the vehicle were studied. The mass distribution of the propellants was approximated with water in the Atlas liquid-oxygen and fuel tanks and the Centaur liquid-oxygen tank. Polystyrene balls were used to simulate the mass of the liquid-hydrogen fuel in Centaur.

Test Setup

All testing described herein was accomplished at the E site of Plum Brook Station. The E site consists of a 140-foot- (42.6-m-) high steel tower designed to accommodate the

Atlas-Centaur vehicle (see fig. 2).

The test vehicle consisted of a complete Atlas-Centaur launch vehicle including nose fairing and Centaur insulation panels. A dynamic model of the Surveyor spacecraft was used in all tests. To facilitate mating with the suspension system, the Atlas was modified by replacing all structure and components aft of the thrust barrel with an equivalent mass beam structure X-frame (see fig. 2). The bending and shear stiffnesses of the vehicle are given in figures 3 and 4 and the mass distribution of the empty vehicle is given in figure 5.

A suspender consisting of a steel cable, a spring box, a hydraulic cylinder, and a load cell was fastened to each of the four frame ends (fig. 2). Each spring box contained 4 to 16 springs with a constant of about 400 pounds per inch (7.0×10^4 N/m) per spring. The number of springs for each test was proportioned to the vehicle weight for the particular configuration to give a static deflection of approximately 1 foot (0.3 m), keeping the natural rocking frequency of the suspension system an order of magnitude below the first bending mode frequency of the vehicle. Lateral stability of the system was provided at the bottom by the $1/2^{\circ}$ (8.7×10^{-3} rad) inclination of the cables and at the top by horizontal springs (fig. 2). The test setup was similar to that used in the longitudinal dynamics tests in reference 1.

To avoid operational problems involved with the handling of cryogenic propellants, an equal volume of deionized water was used instead of liquid oxygen and RP fuel. Polystyrene balls having an equivalent bulk density replaced the liquid hydrogen. With the exception of the first test, the propellant tanks were maintained at flight pressures of 31.0 and 59 psig (21.3×10^4 and 40.6×10^4 N/m^2 gage) in the Atlas liquid-oxygen and fuel tanks, respectively, and 15 and 5 psig (10.3×10^4 and 3.44×10^4 N/m^2 gage) in the Centaur liquid-oxygen and liquid-hydrogen tanks, respectively.

The tank pressures for the first test were maintained at the standby conditions of 17 psig (11.7×10^4 N/m^2) in the Atlas fuel tank, 10 psig (6.89×10^4 N/m^2) in the Atlas liquid-oxygen tank, 10 psig (6.89×10^4 N/m^2) in the Centaur liquid-oxygen tank, and 5 psig (3.44×10^4 N/m^2) in the Centaur fuel tank. The weight and levels of the simulated propellants are given in table I, along with tank pressures used during the tests.

Instrumentation

Instrumentation for all tests consisted of strain-gage-type accelerometers. Locations on the X-axis and Y-axis sides are shown in figure 6. The accelerometers sensed acceleration perpendicular to the longitudinal (Z) axis of the vehicle. In the area of the Centaur insulation panels, holes were cut in the panels to allow attachment of the accelerometers to the tank skin. Load cells were used for both vehicle weighing and measurement of the driving force. All modal data were digitally recorded and reduced with a digital computer program. Sixteen channels of data were recorded on oscillograph paper. All damping was determined by the logarithmic decay method from the analog traces on oscillograph paper. End-to-end system accuracy of the instruments is considered to be approximately 2 percent of full scale. Full-scale ranges of instruments were as follows:

Accelerometers, g (m/sec^2) . . . . ±1.0 (±9.8)
Load cell (force), lb (N) . . . . ±1000 (±4.45×10^3)

Test Procedure

The vehicle tanks (except for the Centaur hydrogen tank) were filled with water and pressurized to simulate the flight conditions as presented in table I. Excitation was then applied to the suspended vehicle by an electrodynamic shaker through a load cell and the X-frame at frequencies varying from 1 to 40 hertz. Input force levels ranged from 30 to 400 pounds (133 to 1779 N).

Inasmuch as a tank rupture hazard existed at the tank pressures used, it was necessary to control all operations remotely once the tanking procedures were begun. These remote operations were conducted from a control room (H-building) located approximately 1/4 mile (400 m) away. Television cameras were used to visually monitor the vehicle.

The resonant frequencies of the vehicle were determined by slowly increasing the excitation frequency while holding the force input constant. Resonance was established when acceleration at the extremities of the vehicle peaked.

When resonant conditions were determined, transducer output was recorded on analog recorders and on digital tape. At the resonance peak, the shaker was electrically decoupled allowing natural decay of the oscillations, with the transducer output being recorded on the analog recorder. This procedure was followed to identify modes for each tanking condition.

RESULTS AND DISCUSSIONS

The test results are discussed in terms of the following major subjects:
(1) Natural frequencies
(2) Mode shapes
(3) Damping
(4) Linearity of response
(5) Phase relations of various parts of the mode

Natural Frequencies

Figure 7 illustrates the variation of natural frequency of each mode with the simulated propellant level associated with vehicle flight times. The variation is shown only for the pitch (Y-Z) plane. Tables II lists natural frequencies as a function of flight time, plane of excitation, and pressure level. It is noted that although tank pods exist in the pitch plane and not in the yaw plane, the effect on natural frequency is minor. Further, two tests run at significantly different pressures show little variation in natural frequency. This is probably due to the linearity of this structure as is discussed later.

Mode Shapes

Mode shapes measured at each of the natural frequencies are plotted in figures 8(a) to (i). The mode shapes are shown for both the pitch (Y-Z) and yaw (X-Z) planes and for two tank pressures for the empty vehicle configurations. Although mode shapes are easily defined as clean, continuous curves in the lower-frequency modes, the higher-frequency modal curves show breaks in the smoothness of the mode shape. A comparison of mode shapes in the two planes indicates that although lower modes compare reasonably, higher modes are significantly different in shape. The shapes of the first two modes in the pitch (Y-Z) plane at two significantly different tank pressures compare reasonably well, as demonstrated by figures 8(b) and (c).

Damping

One of the most difficult parameters to measure in a test of this nature is damping. Even though every effort was made in the design and operation of the test setup to assure that no damping was introduced through the manner in which the vehicle was suspended, no quantitative measure of test-stand-introduced damping was possible. The typical decays shown in figure 9 present an example of the data from which damping was derived. Damping was determined by the logarithmic decay method (ref. 2). A plot of the damping variation with flight time in the Y-Z plane (pitch) is shown in figure 10. A typical value of damping in the pitch plane obtained from flight data during the booster engine cutoff (BECO) transient (T + 150 sec) is also plotted in figure 10. A damping value of 3 percent obtained from flight data compares to 2.2 percent measured in testing at this same simulated flight time.

94

Linearity of Response

One of the more significant problems associated with analysis of control systems and loads interacting with the flexible-body bending modes is the linearity of these modes. Nonlinearities in modal response relative to forcing function level could dictate either convergence or divergence of loads or autopilot stability. Responses of the first four bending modes as a function of force level are shown in figure 11. The data shown in figure 11 indicate the modal response is linear for each of the modes over the excitation range chosen.

Phase Relations of Various Parts of Mode

It is necessary to measure angular position and rate at a number of positions on a typical launch vehicle in order to provide the right amplitude and phase of bending mode relations in the autopilot. In the analysis of these control systems, orthogonal modes are assumed; that is, phase relations between any two parts of a vehicle are assumed to be either 0° or 180° for any vehicle mode, depending on the shape of the bending mode. To evaluate the limitations of this analysis technique, phase relations between force and acceleration were measured for frequencies near each of the first three bending modes. Figure 12 indicates the amount of phase differences associated with each of these frequencies. As the figure shows, very little deviation from orthogonality occurs in the first two modes. However, the third mode indicates significant differences in phase relations.

CONCLUSIONS

Based on the results of the lateral dynamics tests of the Atlas-Centaur-Surveyor vehicle, the following conclusions are drawn:

1. The natural bending mode frequency is only slightly affected by tank pressure.

2. The bending mode frequency is essentially the same for excitation of the vehicle in the XZ and YZ planes.

3. The mode shapes for the first two modes are well defined, while the higher modes become less well defined.

4. Damping of the vehicle averaged about 2 percent of critical.

5. Modal response is indicated to be linear for each of the modes over the excitation range.

6. Very little deviation from orthogonality is indicated in the first two modes. The third and higher modes indicate significant deviation in phase relations.

Lewis Research Center,
    National Aeronautics and Space Administration,
        Cleveland, Ohio, April 8, 1969,
            491-05-00-03-22.

REFERENCES

1. Gerus, Theodore F.; Housely, John A.; and Kusic, George: Atlas-Centaur-Surveyor Longitudinal Dynamics Tests. NASA TM X-1459, 1967.

2. Harris, Cyril M.; and Crede, Charles E.: Shock and Vibration Handbook. Vol. 1. McGraw-Hill Book Co., Inc., 1961.

TABLE I. - TEST CONDITIONS

Test configuration	Units	Test number and simulated flight time					
		I	II and VII	III and VIII	IV	V and IX	VI and X
		Tanks empty	Tanks empty	T + 150 (max. g)	T + 100	T + 75 (max. αQ)	T + 0 (lift-off)
Atlas fuel tank							
Weight of water	lb	0	0	10 050	38 500	51 000	93 000
	N	0	0	44 600	171 000	226 000	412 000
Water level	Station	0	0	1144	1077	1077	940
Tank pressure	psig	17	59	59	59	59	59
	N/m^2	11.68×10^4	40.5×10^4	40.5×10^4	40.5×10^4	40.5×10^4	40.5×10^4
Atlas liquid-oxygen tank							
Weight of water	lb	0	0	14 600	58 300	81 200	150 000
	N	0	0	64 700	258 200	360 000	665 000
Water level	Station	0	0	895	786	734	556
Tank pressure	psig	10.0	31.0	31.0	31.0	31.0	31.0
	N/m^2	6.86×10^4	21.25×10^4	21.25×10^4	21.25×10^4	21.25×10^4	21.25×10^4
Centaur liquid-oxygen tank[a]							
Weight of water	lb	0	0	21 400	21 400	21 400	21 400
	N	0	0	95 000	95 000	95 000	95 000
Water level	Station	0	0	382	382	382	382
Tank pressure	psig	10.3	15.0	15.0	15.0	15.0	15.0
	N/m^2	6.86×10^4	10.3×10^4	10.3×10^4	10.3×10^4	10.3×10^4	10.3×10^4
Total weight (including weight of tanks)	lb	25 530	25 530	71 580	143 730	179 130	289 930
	N	113 200	113 200	318 000	636 000	795 000	1 285 000

[a]Conditions of Centaur liquid-hydrogen tank were the same for all tests: weight of polystyrene, 4630 pounds (20 500 N); level of polystyrene, station 209; tank pressure, 5.0 psig (3.43×10^4 N/m^2).

TABLE II. - EXPERIMENTAL NATURAL FREQUENCIES AND DAMPING RATIOS FOR TEST SIMULATED FLIGHT CONDITIONS

(a) Excitation force in Y-Z plane

Test condition	Mode	Frequency, Hz	Damping ratio
Oxidizer and fuel tanks empty and at standby pressure	1	6.21	0.022
	2	14.20	.024
Oxidizer and fuel tanks empty at nominal flight pressures	1	6.12	0.023
	2	14.51	.053
	3	31.90	.023
	4	37.00	.022
Simulated flight conditions at T + 150 seconds	1	4.93	0.022
	2	9.89	.033
	3	25.71	.014
	4	35.09	No data
Simulated flight conditions at T + 100 seconds	1	3.58	0.025
	2	8.13	.011
	3	17.42	.012
	4	23.12	.009
Simulated flight conditions at T + 75 seconds	1	3.00	0.024
	2	7.77	.013
	3	13.44	.012
	4	20.37	.031
	5	31.44	.002
Simulated conditions at T + 0 (lift-off)	1	2.59	0.019
	2	6.49	.013
	3	9.82	.019
	4	12.74	No data
	5	17.98	0.004

(b) Excitation force in X-Z plane

Test condition	Mode	Frequency, Hz	Damping ratio
Oxidizer and fuel tanks empty at flight pressures	1	6.21	0.037
	2	13.49	.058
	3	31.40	No data
Simulated flight conditions at T + 150 seconds	1	4.74	0.035
	2	10.78	No data
	3	26.62	0.025
	4	36.60	No data
Simulated flight conditions at T + 75 seconds	1	2.97	0.026
	2	8.01	No data
	3	13.35	0.014
	4	20.60	.006
	5	32.89	.016
Simulated conditions at T + 0 (lift-off)	1	2.60	0.023
	2	6.52	.021
	3	10.01	.022
	4	13.01	.026
	5	17.90	.006

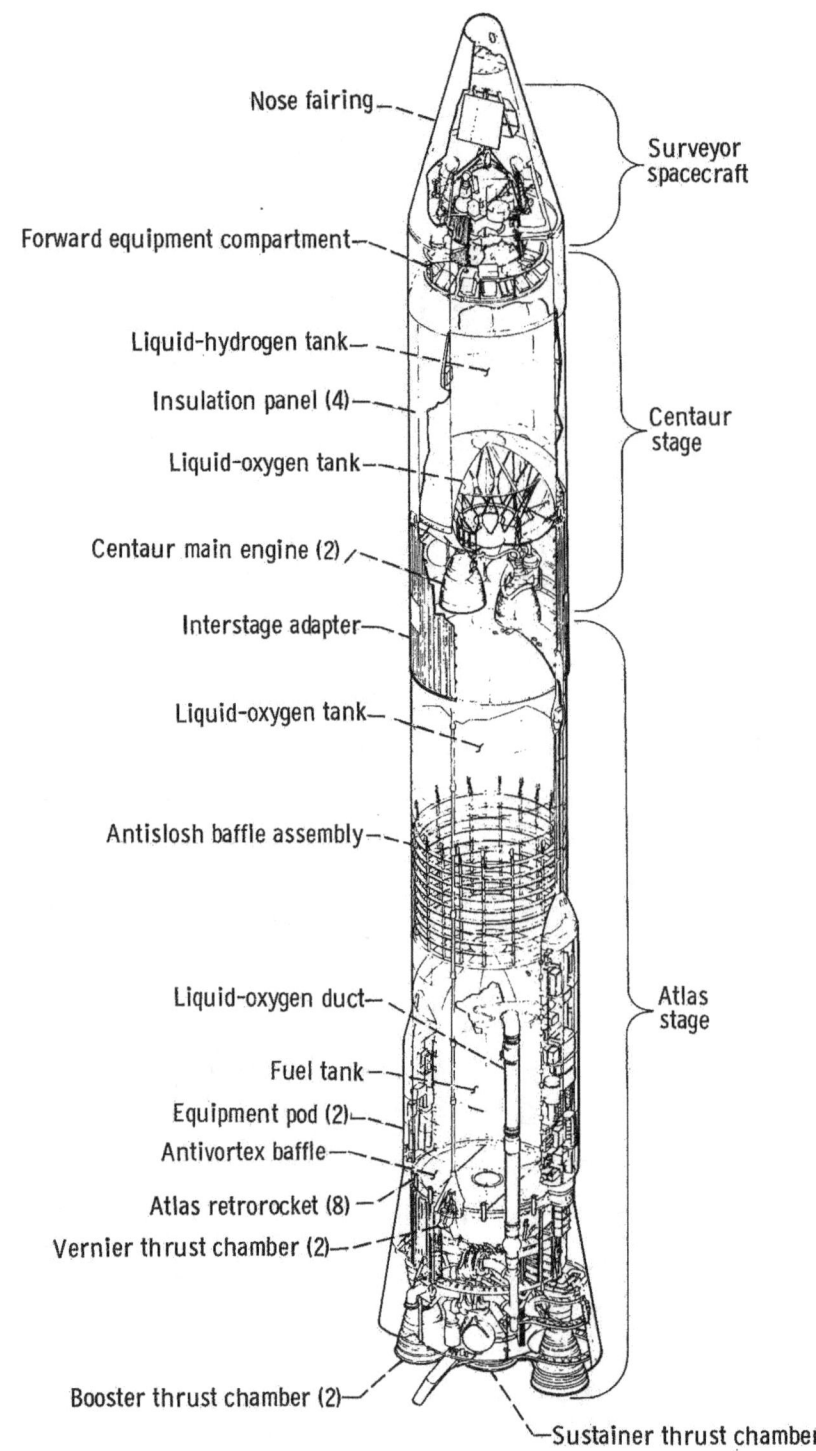

Nose fairing

Forward equipment compartment

Liquid-hydrogen tank

Insulation panel (4)

Liquid-oxygen tank

Centaur main engine (2)

Interstage adapter

Liquid-oxygen tank

Antislosh baffle assembly

Liquid-oxygen duct

Fuel tank

Equipment pod (2)

Antivortex baffle

Atlas retrorocket (8)

Vernier thrust chamber (2)

Booster thrust chamber (2)

Sustainer thrust chamber

Surveyor
spacecraft

Centaur
stage

Atlas
stage

CD-10160-31

Figure 1. - Atlas-Centaur-Surveyor space vehicle configuration.

99

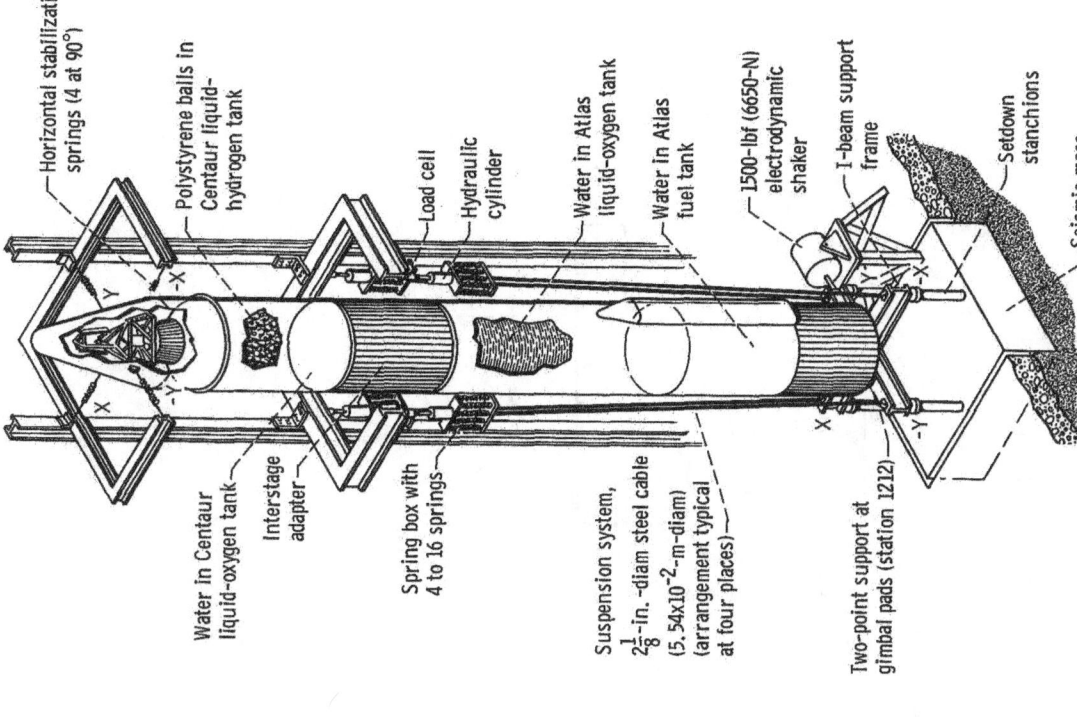

Horizontal stabilization springs (4 at 90°)

Polystyrene balls in Centaur liquid-hydrogen tank

Load cell

Hydraulic cylinder

Water in Atlas liquid-oxygen tank

Water in Atlas fuel tank

1500-lbf (6650-N) electrodynamic shaker

I-beam support frame

Setdown stanchions

Seismic mass

Water in Centaur liquid-oxygen tank

Interstage adapter

Spring box with 4 to 16 springs

Suspension system, $2\frac{1}{8}$-in.-diam steel cable (5.54×10^{-2}-m-diam) (arrangement typical at four places)

Two-point support at gimbal pads (station 1212)

CD-10404-32

(b) Vehicle support system.

Figure 2. – Test setup.

Erection cranes

Interstage adapter

Suspension cable

Atlas

X-frame

Shaker

Nose fairing

Centaur

Deionized water tank

P64-1171.

(a) E-stand test facility with Atlas-Centaur-Surveyor.

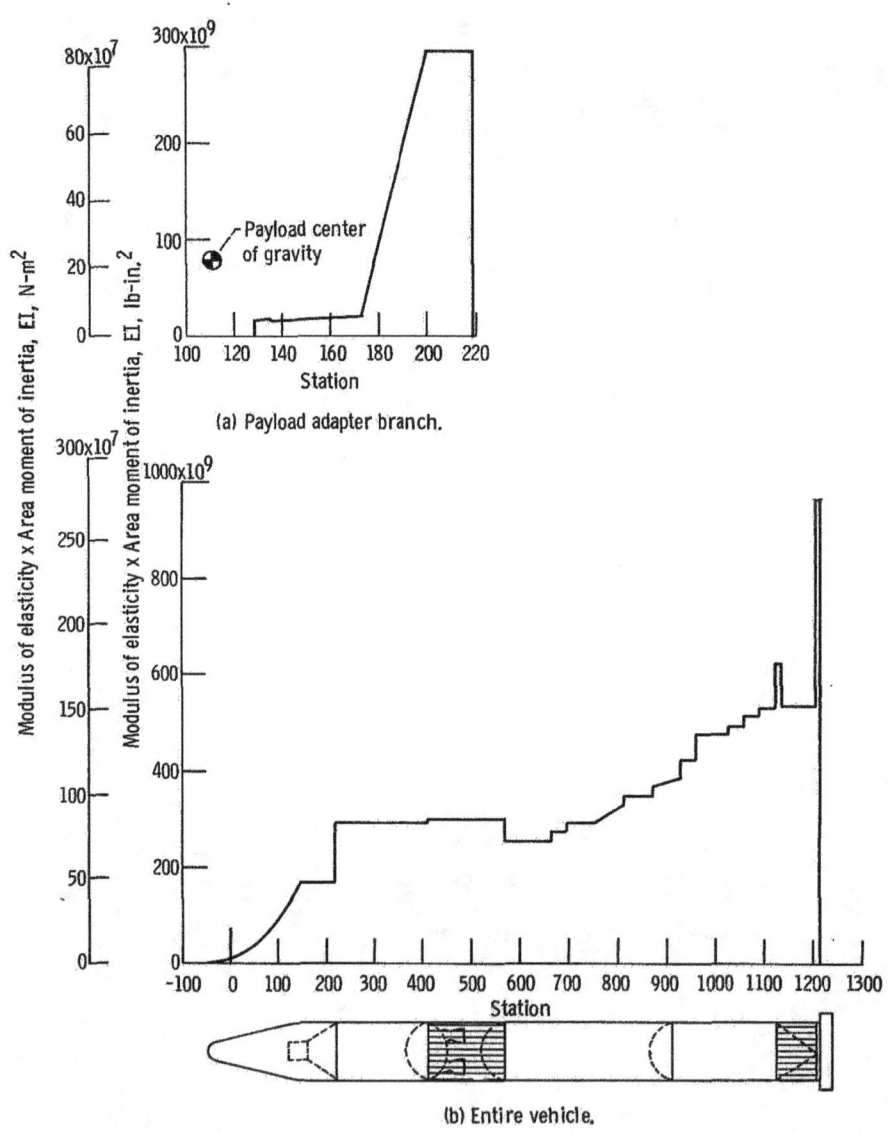

(a) Payload adapter branch.

(b) Entire vehicle.

Figure 3. – Bending stiffness distribution for Plum Brook vehicle.

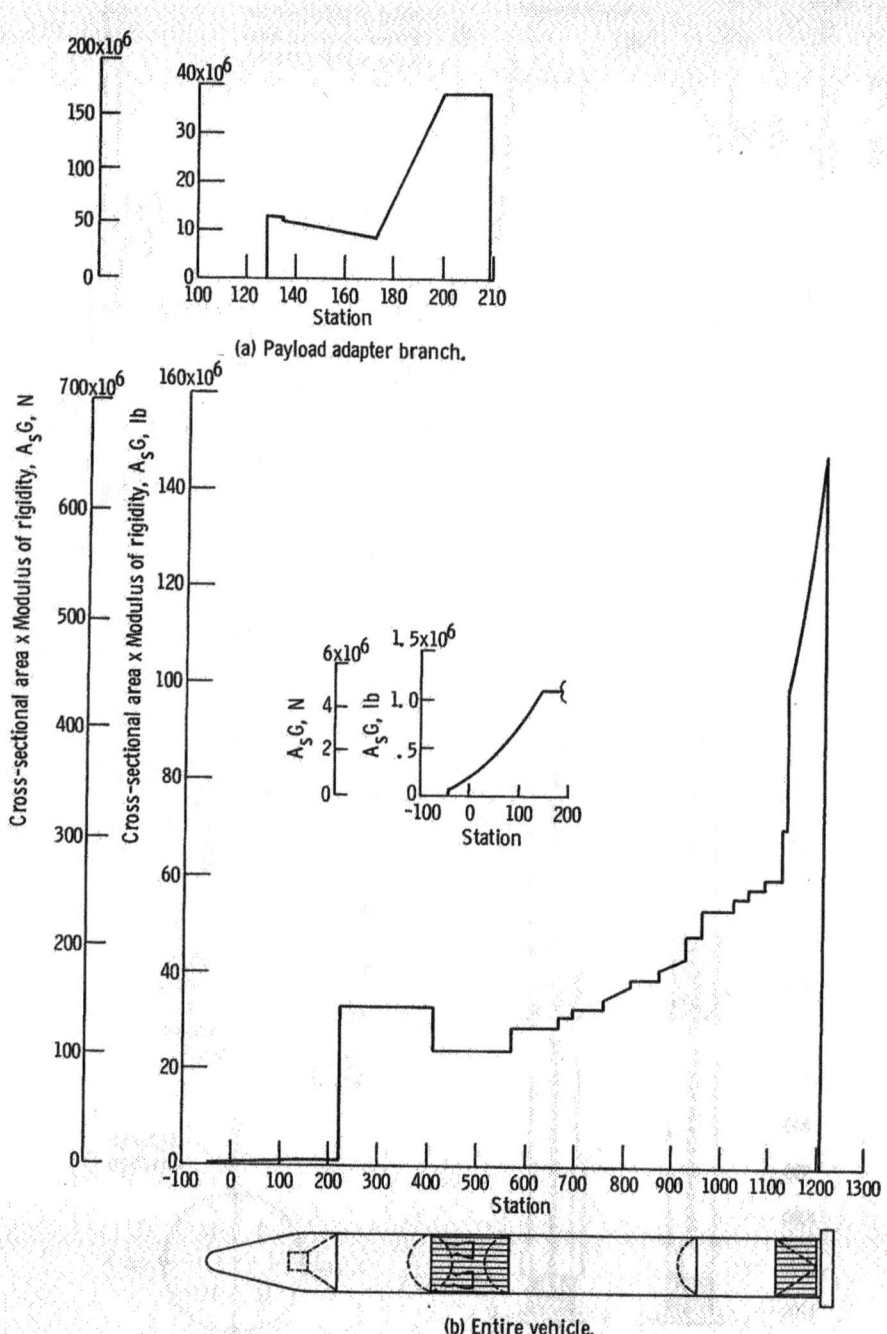

(a) Payload adapter branch.

(b) Entire vehicle.

Figure 4. - Shearing stiffness distribution for Plum Brook vehicle.

102

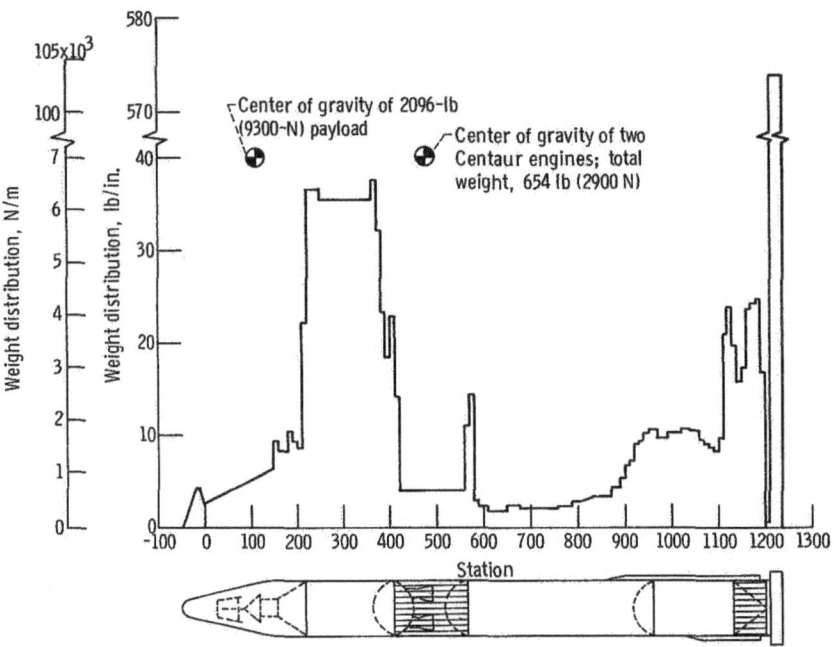

Figure 5. – Weight distribution as function of station for test I configuration.

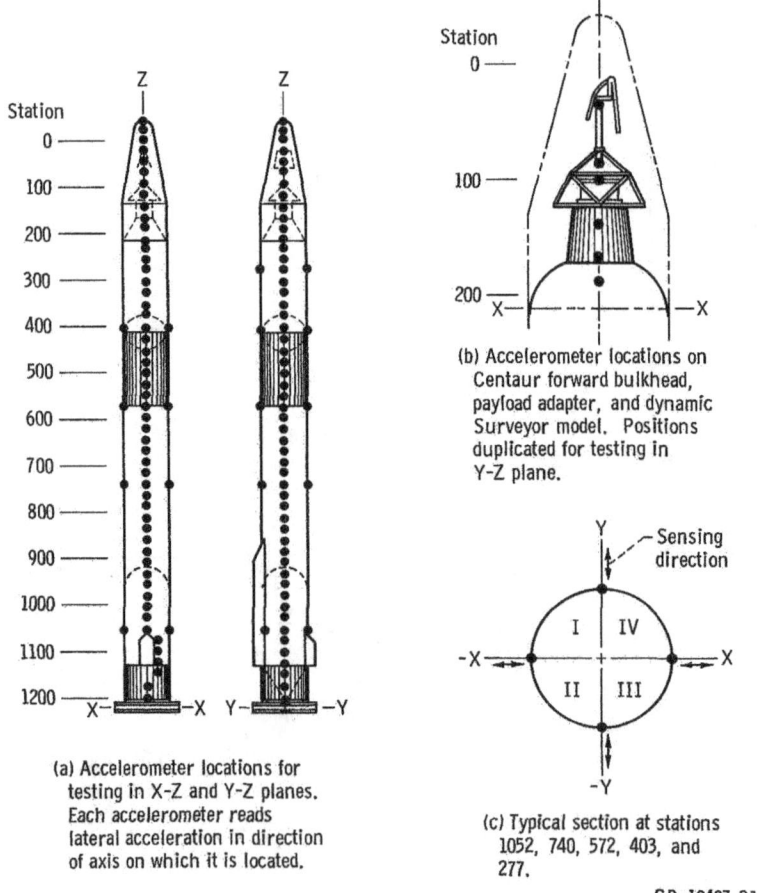

(a) Accelerometer locations for testing in X-Z and Y-Z planes. Each accelerometer reads lateral acceleration in direction of axis on which it is located.

(b) Accelerometer locations on Centaur forward bulkhead, payload adapter, and dynamic Surveyor model. Positions duplicated for testing in Y-Z plane.

(c) Typical section at stations 1052, 740, 572, 403, and 277.

CD-10407-31

Figure 6. – Location of accelerometers.

103

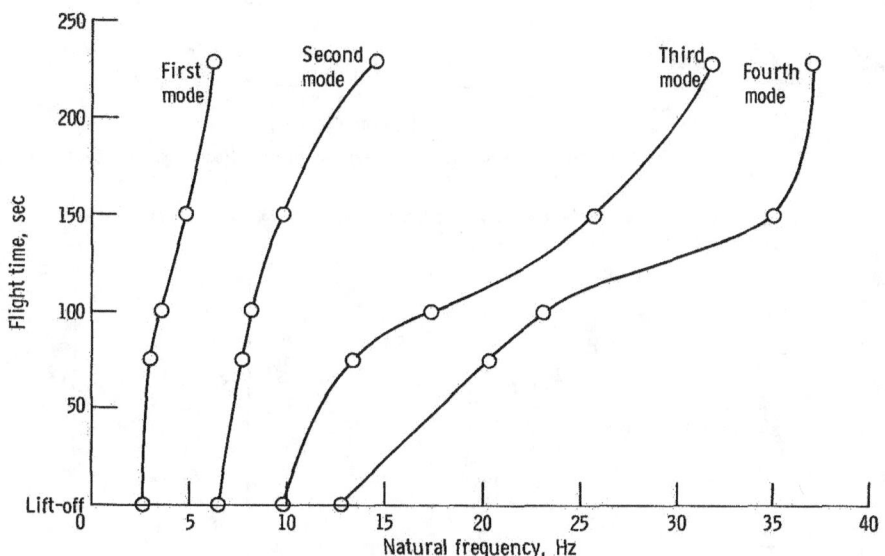

Figure 7. – Variation of natural frequency as function of simulated flight time.

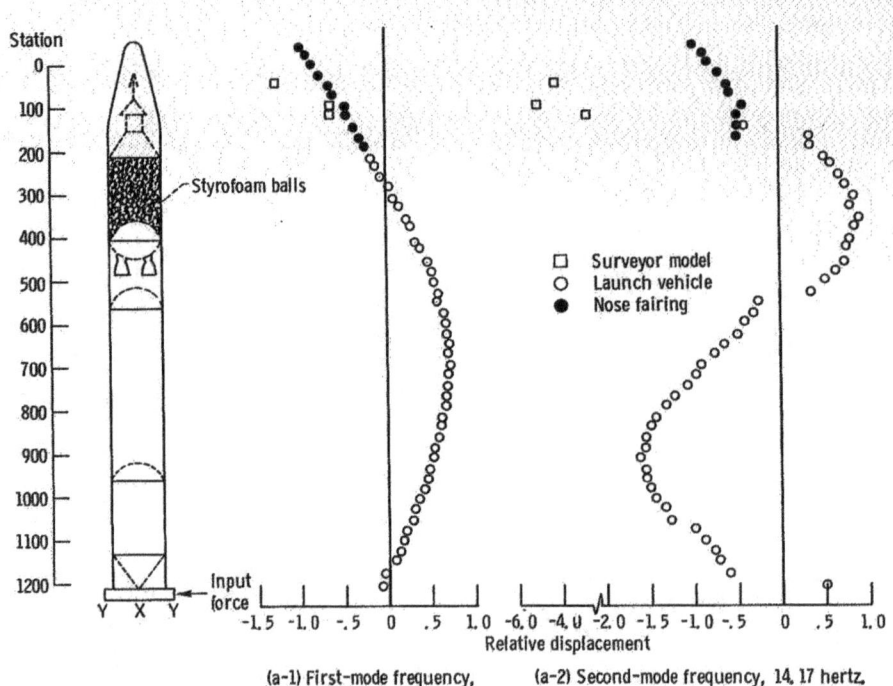

(a-1) First-mode frequency,   (a-2) Second-mode frequency, 14. 17 hertz.
5. 95 hertz.

(a) Oxidizer and fuel tanks empty at standby pressures (10 and 17 psig (6. 89x10^4 and 11. 7x10^4 N/.m^2 gage), respectively); Y-Z plane.

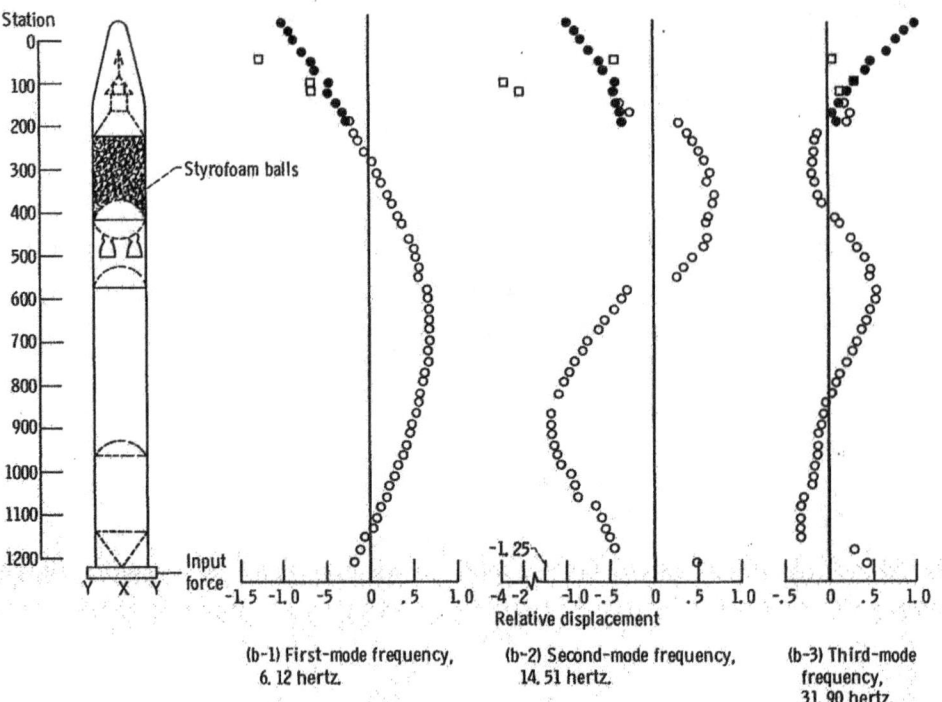

(b-1) First-mode frequency,   (b-2) Second-mode frequency,   (b-3) Third-mode
6. 12 hertz.   14. 51 hertz.   frequency,
31. 90 hertz.

(b) Oxidizer and fuel tanks empty at flight pressure; Y-Z plane.

Figure 8. – Lateral dynamics – bending mode shapes. Surveyor dynamic model SD-4.

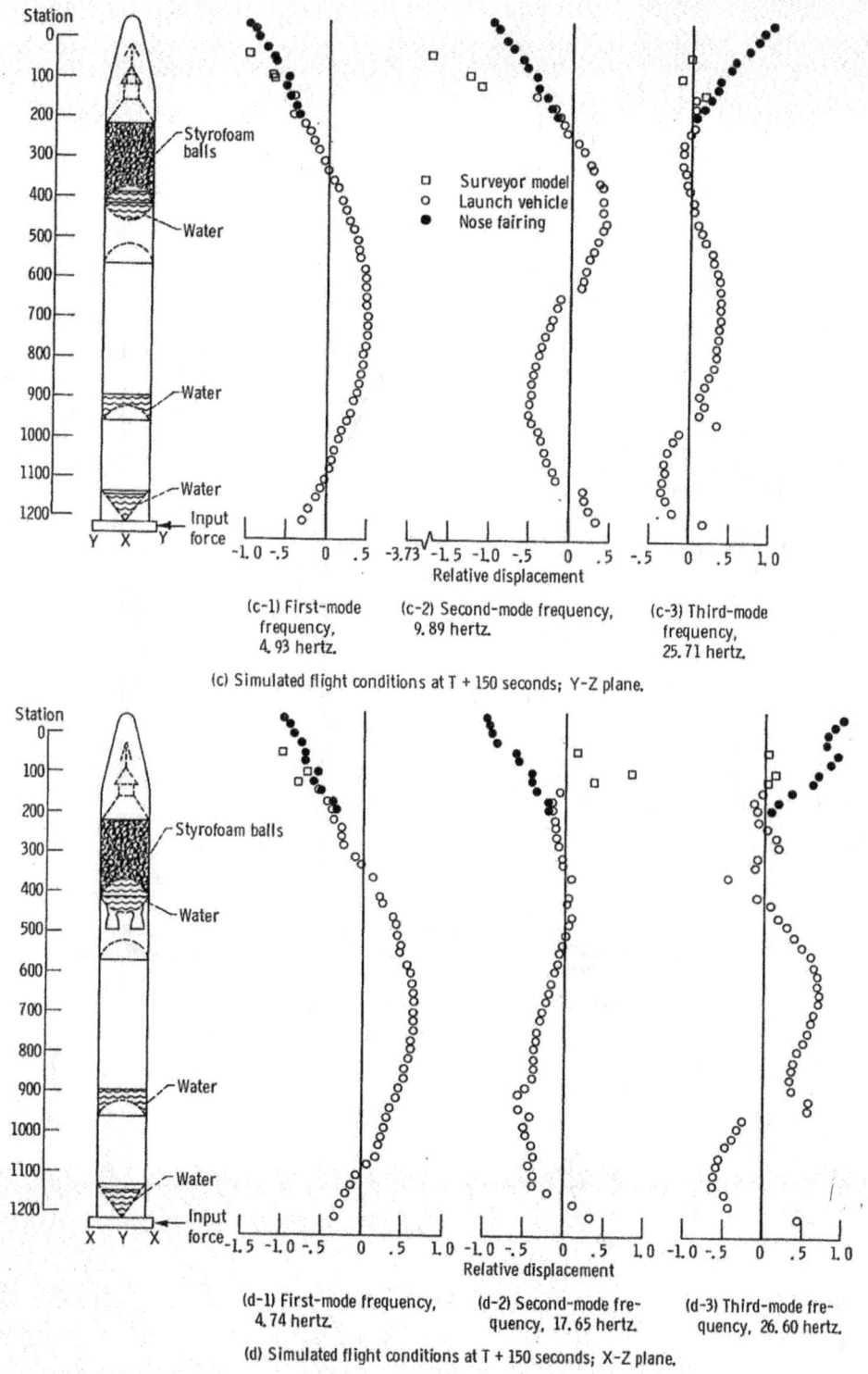

(c-1) First-mode frequency, 4.93 hertz.

(c-2) Second-mode frequency, 9.89 hertz.

(c-3) Third-mode frequency, 25.71 hertz.

(c) Simulated flight conditions at T + 150 seconds; Y-Z plane.

(d-1) First-mode frequency, 4.74 hertz.

(d-2) Second-mode frequency, 17.65 hertz.

(d-3) Third-mode frequency, 26.60 hertz.

(d) Simulated flight conditions at T + 150 seconds; X-Z plane.

Figure 8. - Continued.

106

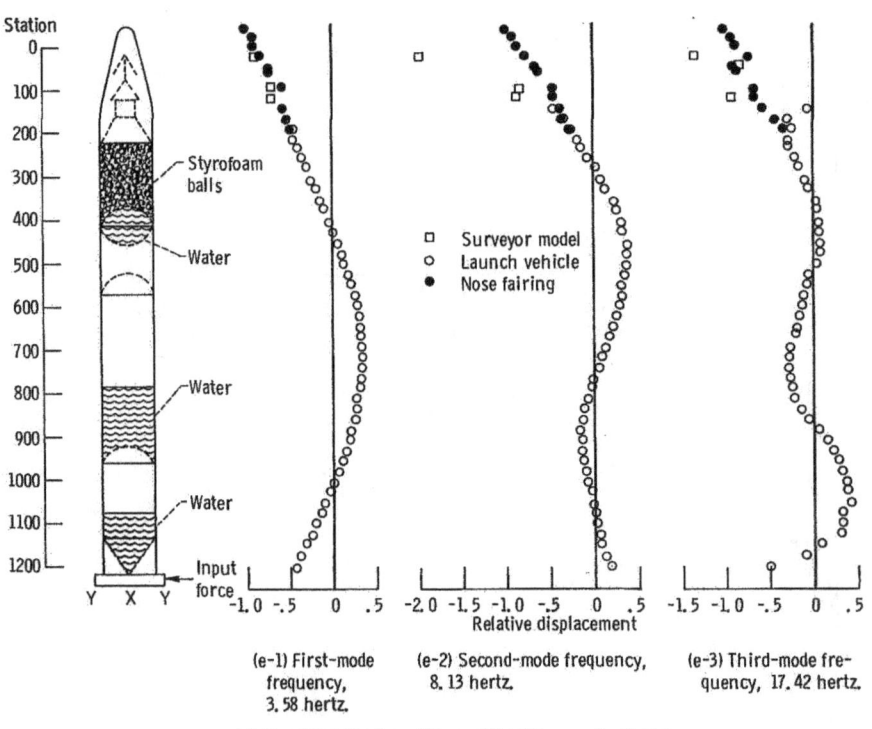

(e-1) First-mode frequency, 3.58 hertz.

(e-2) Second-mode frequency, 8.13 hertz.

(e-3) Third-mode frequency, 17.42 hertz.

(e) Simulated flight conditions at T + 100 seconds; Y-Z plane.

Figure 8. - Continued.

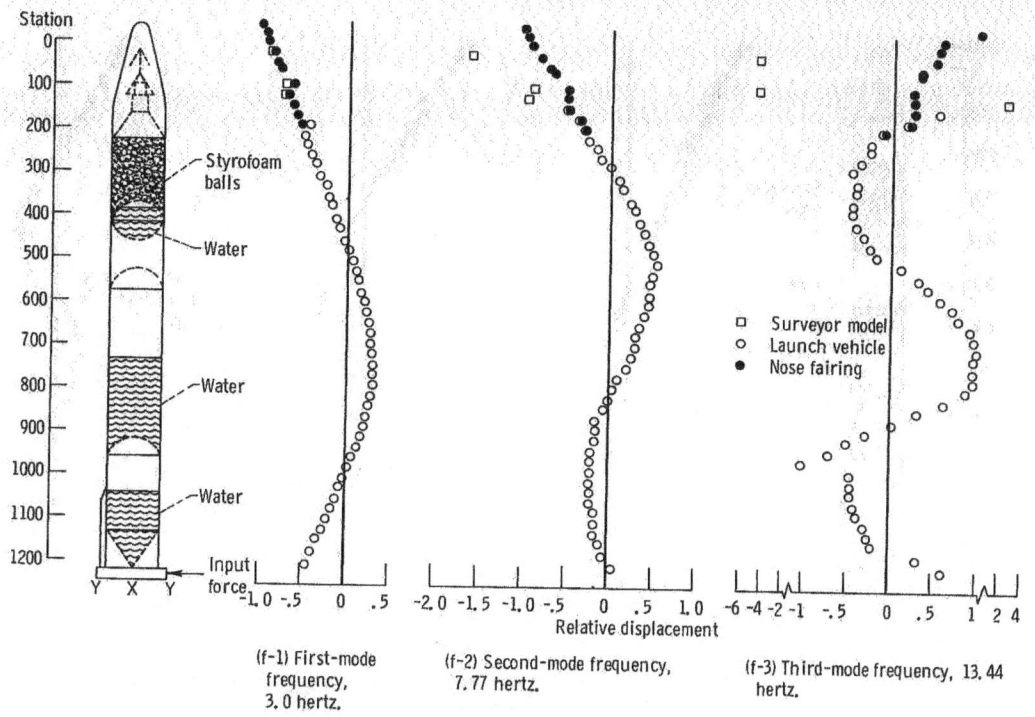

(f-1) First-mode frequency, 3.0 hertz.

(f-2) Second-mode frequency, 7.77 hertz.

(f-3) Third-mode frequency, 13.44 hertz.

(f) Simulated flight conditions at T + 75 seconds; Y-Z plane.

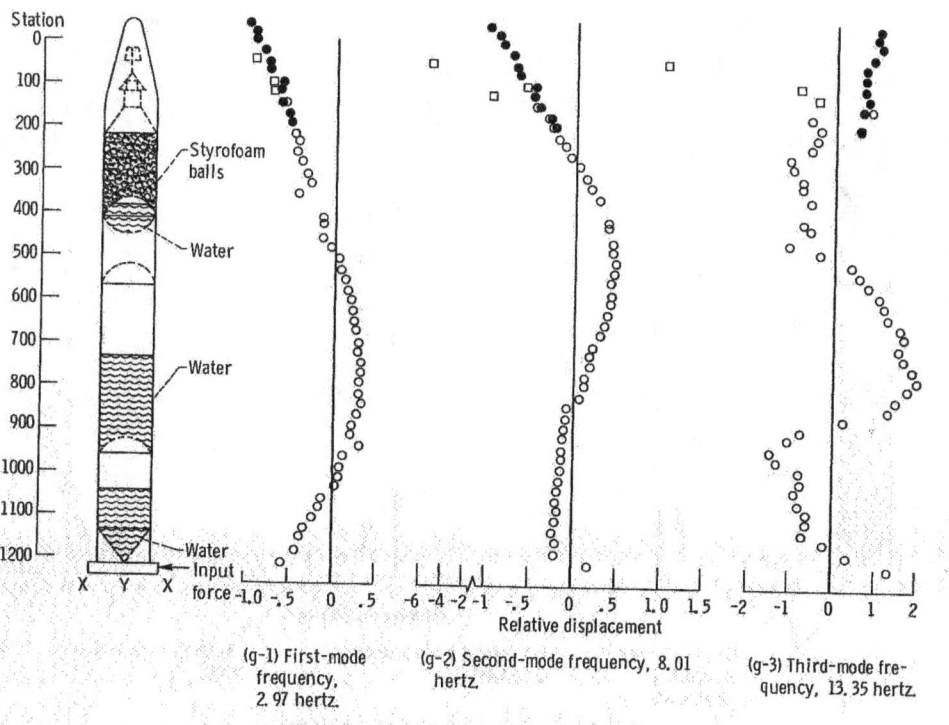

(g-1) First-mode frequency, 2.97 hertz.

(g-2) Second-mode frequency, 8.01 hertz.

(g-3) Third-mode frequency, 13.35 hertz.

(g) Simulated flight conditions at T + 75 seconds; X-Z plane.

Figure 8. - Continued.

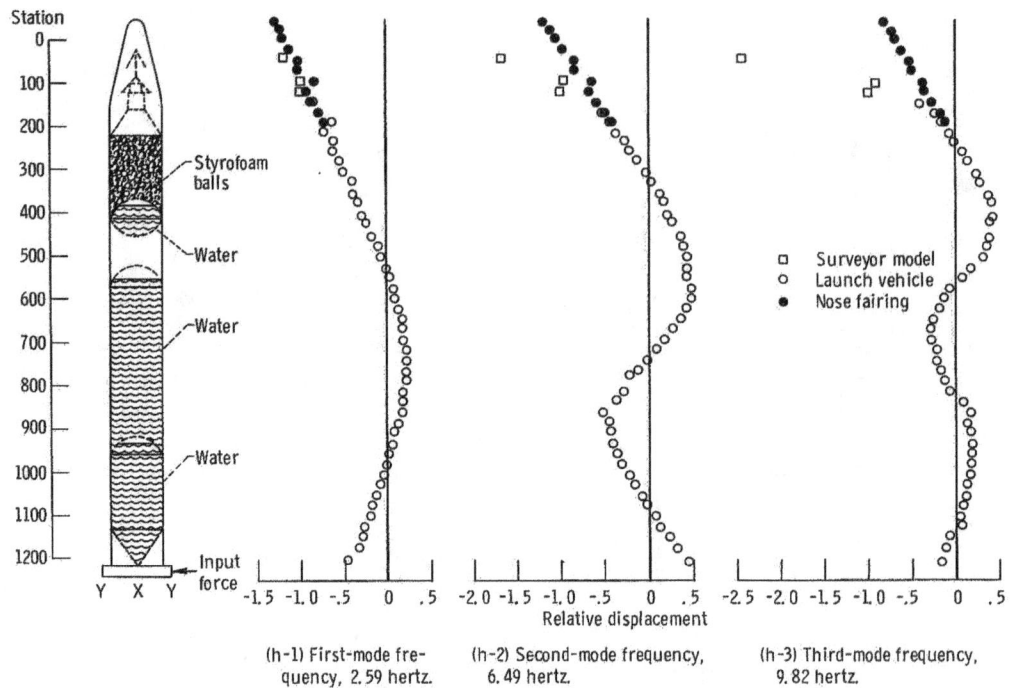

(h-1) First-mode fre-
quency, 2.59 hertz.

(h-2) Second-mode frequency,
6.49 hertz.

(h-3) Third-mode frequency,
9.82 hertz.

(h) Simulated flight conditions at lift-off; Y-Z plane.

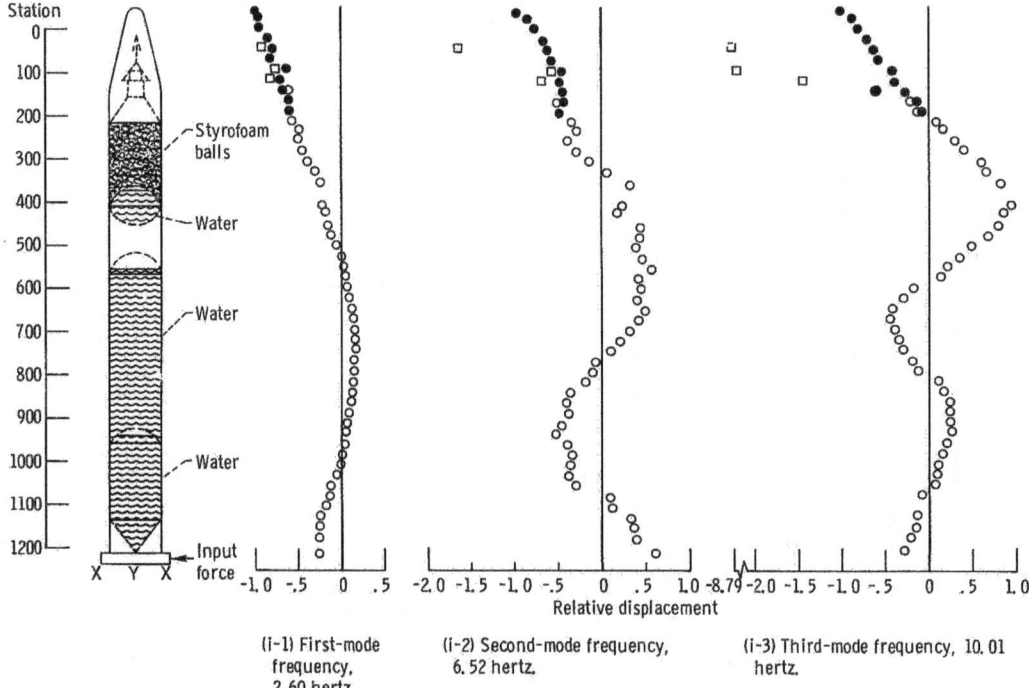

(i-1) First-mode
frequency,
2.60 hertz.

(i-2) Second-mode frequency,
6.52 hertz.

(i-3) Third-mode frequency, 10.01
hertz.

(i) Simulated flight conditions at lift-off; X-Z plane.

Figure 8. - Concluded.

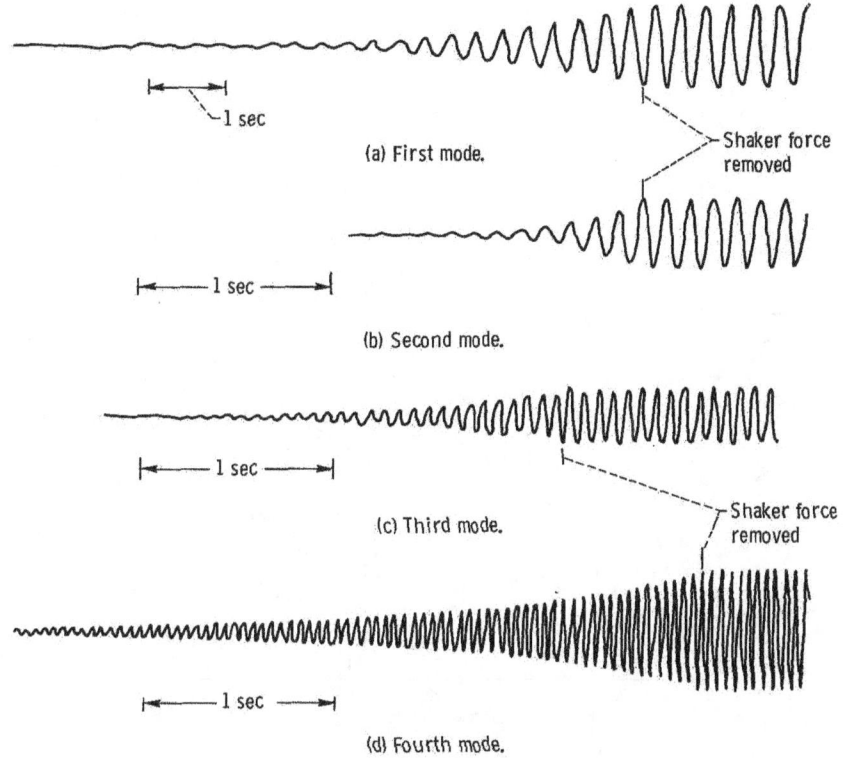

(a) First mode.

~1 sec

Shaker force removed

(b) Second mode.

1 sec

(c) Third mode.

1 sec

Shaker force removed

(d) Fourth mode.

1 sec

Figure 9. - Typical data used to calculate damping by decay methods in Atlas-Centaur-
Surveyor lateral dynamics tests. Simulated flight conditions at T +75 seconds.

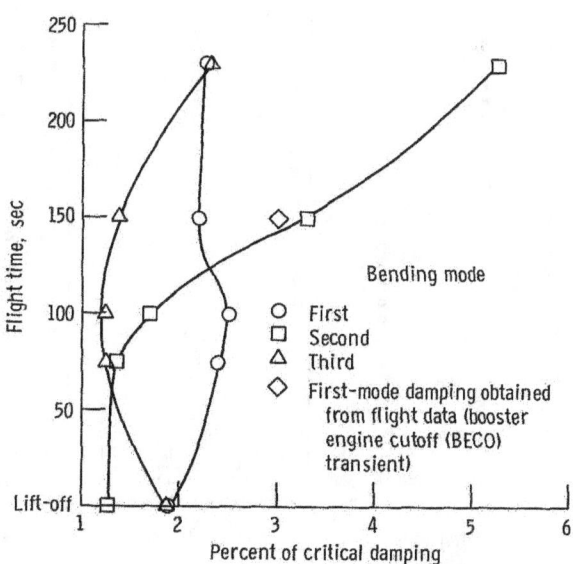

Bending mode

○ First
□ Second
△ Third
◇ First-mode damping obtained
   from flight data (booster
   engine cutoff (BECO)
   transient)

Flight time, sec

Percent of critical damping

Figure 10. - Percent of critical damping as function of
flight time; Y-Z plane (pitch).

110

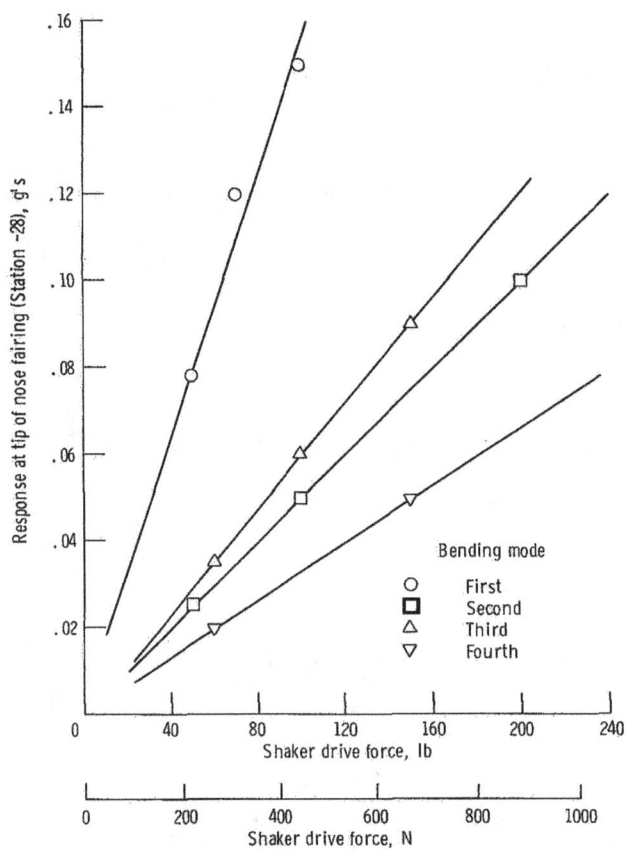

Figure 11. - Response at tip of nose fairing as function of shaker drive force. Simulated flight conditions at T + 75 seconds.

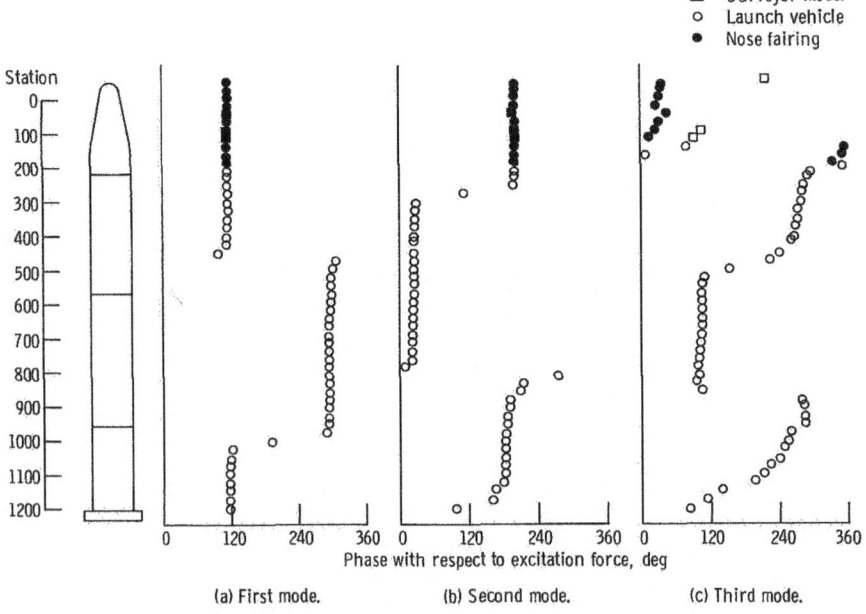

(a) First mode.   (b) Second mode.   (c) Third mode.

Figure 12. - Phase with respect to excitation force as function of vehicle station. Simulated flight conditions at T + 75 seconds; Y-Z plane.

111

Convair Team Returns After 2½ yrs. at Lewis

Personnel from Convair's Dept. 977 have "returned home" from NASA's Lewis Research Center and its Plum Brook Station in Ohio after 2½ years of assisting with tests of Atlas and Centaur vehicles and other space hardware under simulated flight conditions.

Grant L. Hansen, Convair vice president for launch vehicle programs, has received a letter from Edmund R. Jonash, Centaur project acting manager for NASA, commending the Dept. 977 personnel for their services.

Among tests on which they provided assistance were:

Two full-scale inter-stage adapter separation tests for AC-4 and AC-6 configurations.

Structural and dynamic tests with Atlas 116-D.

Several simulated Centaur 6-A flights in a space environmental test chamber and Centaur dual vent valve tests.

OAO nose fairing tests at 100,000-foot simulated altitude.

Surveyor nose fairing tests for AC-4 and AC-6 configurations at 400,000-foot simulated altitude.

Testing of the combined Atlas 116-D, Centaur 5-C and Surveyor dynamic model in E Stands, a dynamic research facility.

Lewis Research Center occupies a 350-acre site in Cleveland and Plum Brook Station is 60 miles distant near Sandusky, Ohio.

J. E. Lauen, chief of Dept. 977's Lewis Research Center test operations and support, said the Convair employes during the 2½ years in Ohio had a high level of morale and efficiency and excellent cooperation from NASA employes with whom they worked.

"One of our primary duties was acquainting NASA personnel with our vehicles, test systems and paperwork," he said.

Simple ingenuity was evident in getting valuable information with minimum cost in some of the tests. At one time, toothpicks stuck in modeling clay were used in determining how much clearance — or lack of it — there was when nose fairings were released under high altitude conditions. Aluminum foil cylinders were used to check resiliency of a net before it was used to catch the heavy fairings.

In the Atlas-Centaur-Surveyor tests, water and styrofoam balls were used to simulate weight of fuels in the Atlas and Centaur vehicles. Lateral, longitudinal and torsional tests were conducted on the entire three-part vehicle.

In Atlas-Centaur separation tests, shaped charge and eight retro-rockets were fired and the vehicles separated with "five degrees of freedom" motion at 100,000-foot simulated altitude in the center's 25-foot-diameter space power chamber.

Two hundred thermocouples were used in addition to normal

(Continued on Page 2)

Team Returns From Lewis

(Continued from Page 1)

flight instrumentation in the Centaur 6-A "high altitude" tests. In addition to assisting with installation and checkout of control consoles and support systems and instrumentation, the Convair men assisted with analysis of data and issuance of problem reports.

NASA personnel are continuing post buckling tests on the Atlas 116-D and Centaur 5-C vehicles and environmental testing on Centaur 6-A with some systems updated to AC-8 configuration.

Convair's Dept. 977 personnel were treated to a "going away" party Dec. 14 in Cleveland.

H. F. Eskesen, F. M. Russell and G. E. Kelley have been reassigned to ETR, Tom Jones to the Point Loma Test Site and G. G. Christ, E. R. Appel, J. P. Gimbel, C. W. Goodin and R. E. Masters to CSTS at Kearny Mesa.

Others, all reassigned to the Kearny Mesa plant, are Lauen, R. F. Brown, H. W. Eger, P. I. Ferguson, J. E. Nixon, F. M. Russell, N. M. Skow, E. H. Vossen and R. E. West.

After the successful AC-4 flight in 1964, John H. Povolny, chief of the test engineering branch at Lewis, in a memorandum to Lauen called successful performance and jettisoning of the Centaur nose cone during the flight a "direct result" of "beyond the call of duty" efforts of men in the department.

Lauen said work of the department also had a bearing on NASA personnel at the E Stand at Plum Brook Station receiving NASA's Group Achievement Award in 1964, the first time for the award to be given to a Lewis group.

Article from the January 26, 1966 issue of *General Dynamics News* (San Diego Edition, Vol. 19, No. 2).

'Site' Building Active At Plum Brook

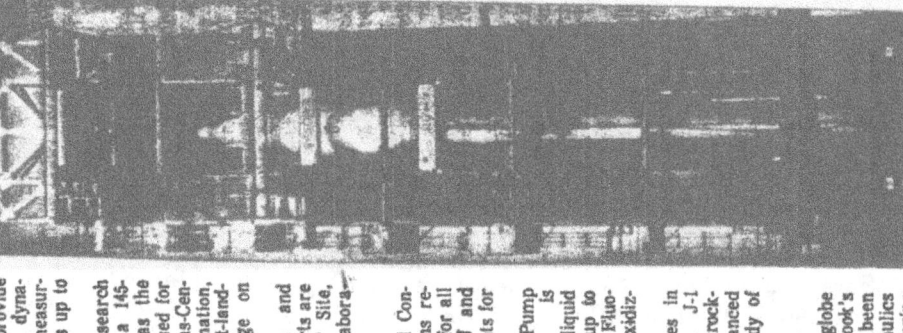

ATLAS-CENTAUR test vehicle is shown in a dynamic test stand at the National Aeronautics and Space Administration Plum Brook Station near Sandusky.

EDITOR'S NOTE: This is the eighth in a series of articles by staff writer Rita Tessmann on research programs being conducted in the Register area. It is also the fourth of a five-part series on the National Aeronautics and Space Administration's Plum Brook Station.

Scattered all over a certain 6,000 acres of partly wooded land in Perkins Township are unusual-looking tall, skinny buildings and long, low buildings that occasionally shake, rumble or shoot out flames.

THEY ARE the Rocket Research Facilities of the National Aeronautics and Space Administration's Plum Brook Station.

Rocket firings at Plum Brook do not involve any actual launching. Hot firings generally involve only the rocket engines which are firmly secured.

To hold these experimental rockets and large parts of rockets or engine components, many structures have been and still are being built. They are known to the men who work there by site designations.

A SITE, B Sites, C Site, and so on are their names. There are 15 sites in all, either existing or under construction.

The Pump Test Area — A Site — houses two large "loops" through which liquified gases at very low temperatures are flowed at a high rate of speed.

Test pumps are inserted into these fluid transport "loops" to determine their performance under the (rigorous conditions of fast flows and a temperature about 300 to 400 degrees below zero Farenheit.

THERE ARE three B Sites. B-2 and B-3 are new construction projects. In the early construction stage, B-2 is an environmental test stand for firing complete upper - stage rocket vehicles under simulated space conditions.

Test equipment is being installed in the most-complete B-3 Site, a 200-foot high stand to be used for non-nuclear tests of various components of large nuclear rocket engines such as will be needed for interplanetary travel.

B-1 is the Nuclear Rocket Engine Dynamics stand. It is also known as the NERVA Stand. NERVA is short for "nuclear engine for rocket vehicle application."

In B-1, 15-second to three-minute tests are being conducted on the propellant system start-up characteristics of the NERVA.

C Site, Turbo Pump Test Area, is equipped to run two simultaneous experiments on pumps. Liquid hydrogen is used in these tests.

TO TEST turbines, D Site has gas generators to provide hot working fluids and dynamometers capable of measuring power from turbines up to 15,000 horsepower.

E Site, Dynamics Research Test Center, features a 145-foot structure known as the "Shake Tower." It is used for ground tests of the Atlas-Centaur - Surveyor combination, slated to place a soft-landing instrument package on the moon.

Fluid flow research and studies of component parts are being carried on at F Site, the Hydraulic Flow Laboratory.

H SITE is the Central Control Building. It contains remote operating controls for all but two sites, B and J and data recording instruments for all sites.

The Liquid Fluorine Pump Laboratory, I Site, is equipped to handle liquid fluorine flows at rates up to 50 pounds per second. Fluorine is the most active oxidizer known.

There are four J Sites in the Rocket Test Area. J-1 and 2 are both hot firing rocket stands. J-3 is an advanced tank test facility for study of space flight fuel storage. J-4 has been abolished.

J-5 is a large steel globe left over from Plum Brook's Army days, that has been converted into a hydraulics laboratory for liquid fluorine compatibility tests.

Heading up Rocket Systems Research is Division Chief Glen Hennings. Hennings has been associated with the Lewis Research Center since 1944.

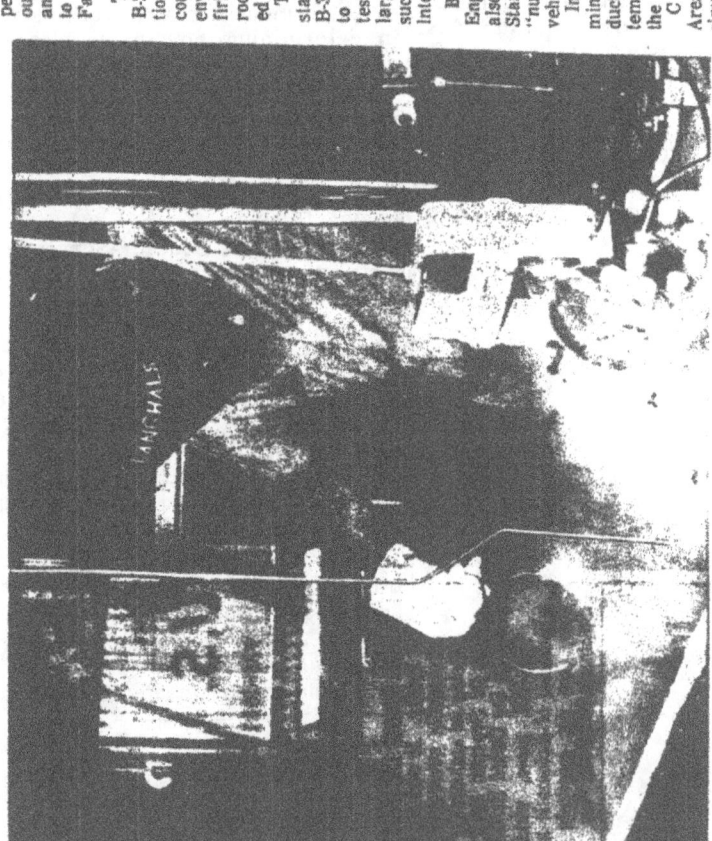

PREPARATIONS for a static rocket firing at NASA's Plum Brook Station are shown here. Technicians are following the flow of liquid nitrogen over the liquid oxygen propellant tanks for the rocket engine to be tested.

Article from the August 14, 1965, issue of *The Sandusky Register*.

113

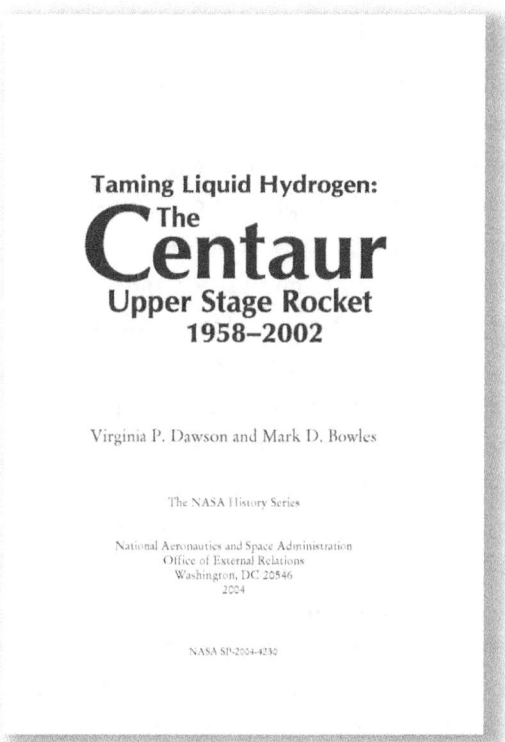

Taming Liquid Hydrogen:
The Centaur
Upper Stage Rocket
1958–2002

Virginia P. Dawson and Mark D. Bowles

The NASA History Series

National Aeronautics and Space Administration
Office of External Relations
Washington, DC 20546
2004

NASA SP-2004-4230

An excerpt from Taming Liquid Hydrogen: The Centaur Upper Stage Rocket 1958-2002 (SP-2004-4230) by Dawson and Bowles, 2004 (pages 71-75):

"Abe's Commandment"

Silverstein's emphasis on ground testing represented a shift from the relatively hands-off approach of the Air Force. Greater supervision by NASA initially produced friction, especially since engineers at General Dynamics were far more experienced in rocket development than their Lewis counterparts. Grant Hansen, head of Centaur for General Dynamics, questioned his new Lewis managers' "willingness to test until everything was absolutely locked down, without any particular regard for an ultimate end date that had to be met." Deane Davis called testing "Abe's commandment." In order to enable General Dynamics to check the interface between Atlas and Centaur before shipment to the Cape, NASA now agreed to the construction of a ten-story, $6.8-million facility adjacent to the Kearny Mesa Industrial Park, despite the fact that General Dynamics already had an impressive array of test facilities.

Silverstein insisted that Centaur also be tested at Lewis. Testing began with components, followed by engine tests, then tests of the entire system, and, finally, full static tests of the entire flight configuration. Separation tests, shake tests, structural tests, nose-fairing tests, and insulation-panel tests all contributed to growing confidence that the Center had the facilities and expertise to monitor the contractor.[32] In Silverstein's view, even if there were only a one-in-a-thousand chance of failure of one component, it was better to test it than to risk failure. His technical perfectionism was legendary. To Silverstein, space vehicles required "not statistical accuracy but one hundred percent accuracy."[33] This could only be achieved by "extreme diligence and the concept that every piece of equipment that is taken aboard, every component, must be proven—environmentally checked so that it can live in the environment of space, the total environment of space, the vacuum of space, the temperatures of space." Only after systematic ground testing was completed was he willing to allow Atlas-Centaur to be launched.

Testing at Lewis not only provided independent verification of contractor performance, but also contributed new solutions to problems. The loss of F-1 had occurred within a tenth of a second after hydrogen gas had been vented from an opening at the top of the vehicle. A test program in the 8-by-6-foot supersonic wind tunnel at Lewis revealed that the hydrogen-venting system posed a fire hazard during flight. The solution was to design a vent fin or snout on the nose fairing that extended about 50 inches from the tank, just far enough away to keep the hydrogen gas from igniting along the hot surface of the vehicle.

As General Dynamics people began to see the concrete results of these test programs, respect began to replace the adversarial relationship between General Dynamics and the government.[34] Testing applied to whole systems, as well as individual components. The laboratory hastily converted the Altitude Wind Tunnel into a vacuum chamber in order to test a full-scale operational Centaur vehicle under environmental conditions to simulate spaceflight. Test engineers "soaked" a Centaur in the vacuum chamber to test the separation system by firing the retrorockets in the normal and failure modes. Lewis wind tunnels were also used to test the RL10 engine, while other liquid-hydrogen tests were carried out in the unique Rocket Engine Test Facility at the Center. Although needed for no more than 3 minutes, the flight gyros received 1,000 hours of testing, with the rationale that if the gyros worked for 1,000 hours, they would run several minutes more in space without a problem. Explosive bolts, which could not be tested, were carefully inspected for defects. If, as Lewis engineers often pointed out, there were 1,000 single failure modes on Atlas-Centaur, they were confident that many "bugs" could be uncovered with adequate testing.[35]

Silverstein believed that in addition to providing industry with test facilities, it was important for NASA to build extensive new facilities for rocket tests. Centaur presented special challenges because its engines had to fire in space. With the Centaur carrying one-of-a-kind multimillion-dollar payloads, how could engineers be sure that the Centaur engines would start in a

[32] Philip Geddes, "Centaur: How It Was Put Back on Track," *Aerospace Management* (April, 1964): 24–29.

[33] John Sloop interview with Abe Silverstein, 19 May 1974, NASA Historical Reference Collection.

[34] Personal communication to the author from J. Cary Nettles, 17 June 1999. See also Eugene Kloman, "Centaur," typescript, MFSC archives, 69–70.

[35] Interview with Bruce Lundin by Virginia Dawson, 7 March 2000.

near-vacuum while being subjected at different times to extremely hot and cold temperatures? To answer these questions, NASA appropriated funds for the construction of a unique facility called the Space Propulsion Research Facility, or "B-2," at the Plum Brook Station, located 55 miles from Lewis Research Center in Sandusky, Ohio. At the Plum Brook's "E Site," the Dynamics Research Test Center, an entire Centaur test vehicle could be mated to the Atlas, along with a test model of the Surveyor spacecraft. Important "bending mode tests" performed by Ted Gerus, head of Dynamics, and Robert P. Miller, project lead engineer, involved tanking the vehicle with simulated propellants and shaking it horizontally and vertically to test its structural integrity. The Atlas also received axial load tests at 248,000 pounds to simulate the forces on its thin skin during liftoff. It passed with flying colors.

These tests of the steel balloon tank structure provide another example of the cooperation between General Dynamics and Lewis engineers. David Peery, former head of the Aeronautical Engineering Department at Penn State University and author of an important textbook on aircraft structures, had developed a theory for structural strength available after the onset of local skin buckling. At General Dynamics, he showed by analysis that the Atlas-Centaur tank design—thought by many to be too radical—was actually unnecessarily conservative. According to his theory, it could withstand vastly greater stresses. As Richard Martin, author of several excellent articles on Atlas, pointed out, "Peery used his famous bulldog tenacity to convince management at General Dynamics and Lewis to test a full-scale Atlas-Centaur to the point of collapse."[36] Martin explained, "A bending moment about 80 percent greater than that at the outset of buckling was achieved before significant nonlinear deflections occurred, but there were some worrisome local buckles around protuberances like the liquid oxygen line outlet. Therefore, the analysis was adapted to allow only about a 60 percent increase in the applied load."

Collaboration began at this time between Martin and Gerus, leaders of dynamic analysis groups at General Dynamics and Lewis, respectively. These groups laid the foundation for an innovative pitch and yaw program that was later implemented when a new Teledyne digital computer was introduced after 1973. The Automatic Determination and Dissemination of Just Updated Steering Terms (ADDJUST) program allowed design loads from flight winds to be reduced by about 40 percent.

Silverstein insisted on a program review at Lewis once a month. Contractors were expected to spend two or three days at the laboratory. Vince Johnson, Centaur program manager from Headquarters, attended these meetings but was careful not to come between General Dynamics and its new Lewis managers. Johnson had the job of defending the program to Congress and running interference at Headquarters. In Bruce Lundin's view, Johnson "played the role of

---

[36] Communication to the authors by Richard Martin, 7 March 2002. See Richard E. Martin, "The Atlas and Centaur 'Steel Balloon' Tanks: A Legacy of Karel Bossart, "40th International Astronautics Congress paper, IAA-89-738, Cot 7-13m, 1989; "A Brief History of the Atlas Rocket Vehicle, Part III," *Quest—The History of Spaceflight Quarterly* 8 (2001): 48. See also David Peery, *Aircraft Structures* (New York: McGraw-Hill, 1949).

program manager absolutely perfectly."[37] His levelheaded approach to program management was one of the keys to the program's ultimate success.

Deane Davis, General Dynamics Centaur project manager, recalled that General Dynamics grudgingly admitted that Lewis people knew somewhat more than they did about liquid hydrogen. The old Centaur design had required the engines to be prechilled for a short time before Centaur engines were ignited in space. This wasted some of the precious liquid-hydrogen fuel. One of the project's important innovations was to "chill down" the vehicle on the ground, using liquid helium as a precoolant for the engine. This was a direct transfer of knowledge from Project Bee, which included precooling the engine. Liquid helium was a relatively rare chemical at that time, but it was much less difficult to handle than liquid hydrogen. Engineers at General Dynamics were skeptical that it could be procured in large enough quantities. Within minutes of leaving a contentious meeting, Silverstein was on the phone ordering a dewar of liquid helium from the government cryogenics laboratory in Colorado. Liquid helium precooling worked perfectly.[38] General Dynamics and the Lewis engineers also worked together on a new design for the insulation panels—the source of failure on the first launch.

During this time, General Dynamics engineers also solved the critical problem of the leaking of liquid hydrogen through minute pores in the welds of the liquid-hydrogen tank. Tests at General Dynamics revealed that metal became brittle when exposed to the very low temperatures of liquid hydrogen. It was found that by adding nickel in the weld area, the stainless steel could be strengthened. New techniques for lap-welding the seams, followed by spot welds, provided additional structural integrity. Factory technicians, already highly skilled builders of the Atlas tank, had to raise the standard even higher. The vehicle required more than 74,000 spot welds, 360 feet of resistance seam welds, and 400 feet of heliarc fusion butt welds. The welds were carefully monitored and x-rayed to make sure there were no imperfections.[39] In the most critical area of the design—the intermediate bulkhead—where cracks as small as 1/10,000 of an inch could destroy the vacuum created by freezing out the dry nitrogen gas in the space inside the double wall, the gores were fusion-welded.[40] Advanced welding techniques ensured a high degree of structural integrity without adding excessive weight.

Getting to the bottom of the problem of leaks in the intermediate bulkhead led to the discovery of another potential problem. Small cameras placed in the hydrogen tank revealed that as soon as the liquid hydrogen was loaded, the bulkhead mysteriously wrinkled. "Everyone took a look at those pictures and just about fainted, me included," Deane Davis recalled. "And

[37] Interview with Bruce Lundin by Virginia Dawson, 7 March 2000.

[38] John Sloop interview with Aerospace Division, Convair/General Dynamics, 29 April 1974, NASA Historical Reference Collection.

[39] Irwin Stambler, "Centaur," *Space/Aeronautics* (October 1963): 74–75.

[40] Ibid.

of course Abe Silverstein got all upset about it."[41] Testing determined that when the hydrogen was removed, the wrinkles disappeared. Through further testing and analysis, the team concluded that the stainless steel experienced "cryoshock" from the very cold liquid hydrogen. The way to avoid cryoshock was to load the liquid oxygen first, allowing the system to chill down gradually. Then the liquid hydrogen was slowly loaded. Wrinkles in the bulkhead no longer appeared. At the same time that they were solving these problems, engineers at General Dynamics and Lewis were building a formidable expertise in liquid hydrogen that they freely shared with Douglas and Rocketdyne, the contractors for the Saturn upper stages.[42]

---

[41] John Sloop interview with members of the Aerospace Division Convair/General Dynamics, 29 April 1974, NASA Historical Reference Collection.

[42] Roger Bilstein, *Stages to Saturn* (Washington, DC: NASA SP-4206, 1980), 153, 188–189.

America's
Finest
Rockets™

MOONCAT®
COLLECTIBLES
SAN DIEGO, CA

www.ingramcontent.com/pod-product-compliance
Lightning Source LLC
Chambersburg PA
CBHW080301180526
45167CB00006B/2617